MW00450241

HUNTERS OF MAN

True Stories of Man-eaters, Man-killers and Rogues from Southeast Asia

HUNTERS OF MAN
True Stories of Man-eaters, Man-killers and Rogues from Southeast Asia

by
Capt. John H. Brandt (Ret)

with contributions by
Gordon Young and Pat Byrne

Illustrated by
Gordon M. Allen

Safari Press Inc.

HUNTERS OF MAN by Capt. John H. Brandt. Copyright 1989 by Capt. John H. Brandt. All rights reserved. No part of this book may be used or reproduced in any manner whatsoever without prior written permission from the publisher. All inquiries should be addressed to: Safari Press, Inc., P.O. Box 3095, Long Beach, CA 90803, U.S.A. Tel: CA and worldwide (714) 894-9080; U.S.A. (except CA) and Canada (800) 451-4788.

Brandt, Capt. John H.

ISBN 1-57157-018-7

1989, Long Beach, California, U.S.A

10 9 8 7 6 5 4 3 2

Readers wishing to receive the Safari Press catalog, featuring many fine books on big-game hunting, wingshooting, and firearms, should write the publisher at the address given above.

This is the 59th book published by Safari Press.

Other books and publications on Asia by the author

The Asian Hunter

The Southeast Asian Negrito

The Negrito of Peninsular Thailand

Mongolia's Mad Maral

Malaya: Hunting the Seladang

Sambar: Stag of the Indian Shadows

The Wild Cattle of the World

By Dunong and Bouj

Nan Matol: Ancient Venice of Micronesia

Hunting: Southeast Asia and Indo-China

This book is dedicated to Du-Mi, who guided, watched and guarded on dark jungle paths

Contents

Contents

INTRODUCTION

From that distant time, some three million years ago, when our Australopithicine ancestors decided to forego a life in the treetops and try their luck as bipedal, land-dwelling creatures, mankind must have been a constant and easy source of prey for the larger carnivores. Man has not been blessed with sharp canines, claws, great speed or phenomenal strength for an animal our size. As gatherers we could have competed well and judging by the tooth structure of some of the early man-apes, vegetable matter formed the main diet of these varieties. Hunting, as predators ourselves, was only possible by the fact that we could organize group action. We had a brain. We could think. We could plan. Early competitors must have been saber-toothed cats, cave bears, dire wolves and the like. Each, in turn, passed on into history; but man, the most successful of all predators, survived to reach the top of the biological food chain. Although puny by physical standards, man created weapons from rocks, to clubs, to spears, to bows and ultimately to rifles. When considering how recently in historical time this latter innovation came into being, it can be easily realized how long man has had only minor odds in his favor. With the advent of the rifle the situation changed quickly. Yet, in many parts of the world, until the present time, many rural people do not possess such arms to adequately protect themselves.

Most large predators, capable of adding man to their daily menu, gradually gave way. Those that fought back perished. The smarter surviving predators quickly learned man was to be avoided and left alone under normal circumstances. This latter comment is important to keep in mind when discussing animals that kill or eat man. For example, the top predator in North America, the grizzly, was quickly exterminated in areas where he was apt to contest the territory with man. By contrast, the aggressive little barren ground grizzly has not yet learned this lesson. The polar bear has also not yet fully learned that man is not an edible sled-driving seal, and often gets himself killed as a consequence.

In Africa, tales of man-eating lions abound although the history of western man's penetration of that continent is quite recent. Modern weapons quickly brought an end to such depredations. Although killings still occur at frequent intervals, the cats rarely attain much notoriety before someone puts an end to their unpopular practice of killing people—but it was not always so.

Asia, particularly India, is rich in its literature of the descriptive horrors of man-eaters and man-killers. This is largely due to the fact that India had, until quite recently, the largest concentration of large cats in the world—the Bengal tiger and the leopard. At the turn of the century, an estimated 40,000 tiger lived on the Indian sub-continent. No one knows how many leopard shared their environs with the burgeoning human population. Many man-eaters attained tallies of human kills that boggle the mind such as the Champawat tiger that killed 200 victims in Nepal and then added 234 more in India; or the Rudraprayag leopard that killed 125 people, although unofficial numbers are far greater. The Sundrakumpa man-eating tiger with over 400 kills to its credit or the Chowgarh tiger with 64 victims have become classics. Writers such as Jim Corbett, with his excellent stories of the man-eaters of the Kumaon District of northern India during the period prior to World War II, are well-known to many readers. No one can read them without feeling the chilling tension of a sit-up over a human body or the trembling anticipation as a known man-eater approaches the hunter awaiting it.

Even though man-eating and man-killing is largely now viewed in a historical perspective, it is important to note that, although the world will never again see situations where over 850 persons were killed during the first eight years of this century in one district of Bombay alone, man-eating has continued on to a surprisingly large degree to this very day.

In India, it is currently estimated that nearly 100 persons per year are killed by tiger or leopard. In the coastal marshes of the Sunderbans, east of Calcutta and extending into Bangladesh, virtually all tiger are considered as man-eaters and in 1987, a total of 27 people were killed by tiger in this limited area. Statistics for all of India indicate that at least 600 known persons have died from tiger attacks in the past twelve years alone. No, man-killing is not entirely a thing of the past.

In the rest of Asia, from Burma and Thailand to Malaya and Indo-China, similar situations existed but the world had little access to stories from these locales. Stories of man-eaters from Indo-China appeared largely only in French literature and journals. One exception is Malaya, where the former British civil servant, Lt. Col. A. Locke, aptly described his numerous hunts for man-eating tiger in the jungles of Trengannu on the eastern side of the peninsula in the 1950's. At that time there were an estimated 5,000 tiger in Malaya. Now the figure is nearer 500, so the possibilities of interaction between man and predator are considerably lessened. Similarly, in India, by 1972, the tiger population had been reduced to less than 2,000 animals. This low ebb has now been reversed and over 4,000 tiger again live in India. As the tiger population expands in isolated islands of forest, identified as reserves, sanctuaries or parks, surrounded by an ever-expanding population of agriculturists, killings are bound to increase again. Much popular sentiment has been focused on the plight of the declining tiger population, but as soon as man-killings again start to increase, sentimentality will be discarded to the detriment of the tiger.

Animals which dare to attack man, for whatever reason, are quickly earmarked for destruction. Bounties are paid on them and hunters are rewarded and often considered as local or national heroes for having successfully removed such a creature. Well-intentioned conservationists have not made allowances for this eventuality and no meaningful management tool presently exists for controlling or removing surplus animals before attacks on man occur. Until the past decade or two, however, this was not the situation and this is the period that this book is about.

Before developing a book on man-eaters, one must clearly define the terminology used. A man-eater is a predator, usually tiger or leopard, that has developed the abhorrent practice of selecting man as his prey with the grisly intention of consuming him. Often this is brought on by an injury, natural or often inflicted by man, which in some way has incapacitated the cat so it cannot catch its normal prey. Young animals raised by a man-eating parent often learn this practice and continue it even though physically capable of normal hunting. Frequently cats in attacking domestic stock have encounters with herdsmen protecting their animals and quickly learn how vulnerable and easily killed a man can be. It is then often an easy transition from cattle-killer to man-killer and ultimately man-eater.

Man-killing is a slightly different situation in that the ultimate objective is not to eat the victim. Many animals can, and often do, become man-killers. An animal protecting its young may strike and kill someone wandering into the environs of its den. An elephant raiding gardens will sometimes kill people who get in its way or who are attempting to drive it from their crops. Asian elephant in particular seem to develop a vengeful and highly dangerous attitude toward man, particularly in areas where it has not yet learned the hazards of firearms.

Bear, with poor eyesight and often totally engrossed in food gathering, sometimes encounter man in the forest and the bear has been programmed by nature to use offense as the best defense. Capable of inflicting horrible damage, Asian bear, whether black, sloth or sun bear types, are all potential man-killers. Hill people in much of Indo-China have a greater fear of bear attacks than that of tiger or leopard since the attacks are so often totally unprovoked. Outside of India, leopard, as a rule, rarely become man-eaters and attack man only if cornered or provoked. This has surely not been the case in India, however!

Wild boar grow to huge size and also often develop abnormally aggressive tendencies toward men. With razor sharp tusks and incredible speed and ferocity, man-killing boar have, at times, terrorized entire villages before being killed themselves.

Many documented cases of gaur, the largest of the world's wild cattle, along with Asian water buffalo, becoming killers of man, are well-known. Often this is brought on by festering wounds or constant harassment which brings about an abnormal behavior pattern in which the animal chooses to kill rather than avoid and evade man.

With growing populations throughout Asia, many large animals who, in times past, might have lived their entire lives without ever encountering man, are now thrown more and more into frequent contact with each other. Predators who might have normally had an inbred fear of man now are permitted to develop a familiarity with man which often leads to conflict and problems. Leopard have proven to be especially adaptable to living around humans and as long as an understanding exists that prey may include goats, calves and maybe even a pet dog, but never man, it works out as a reasonably acceptable symbiotic relationship. Only if man becomes the target is the situation altered!

In dealing with such man-killers, man-eaters, rogues or marauders, a clear distinction must also be made in the mind of the reader that such removal and destruction need in no way resemble what we commonly think of as sports hunting. In their case, the end justifies any means! Whatever is necessary to destroy the killer or rogue is acceptable—snares, poison, automatic weapons, traps and even group action.

For many years, several of the stories in this book were not written because I, too, thought only in terms of sports hunting and did not include these experiences within that framework. Obviously, they weren't and shouldn't be. This, however, in no way detracts from the excitement and potential danger faced in each such encounter, since the animal has to die because society has deemed that end result necessary.

Since the days of World War II, few hunters have had the extensive experience in dealing with man-eaters and man-killers that outstanding hunters such as Jim Corbett, Kenneth Anderson, or Col. Locke had in the days of the recent past. There were a few notable exceptions, however.

I spent eleven years in Asia, primarily on the Malayan Peninsula, Thailand, and in the Pacific Islands. I have combined a few experiences of my own with those of two eminent colleagues with whom I spent much time in the field and in pursuit of animals described in this book.

Gordon Young, author of *Tracks of an Intruder* and *Hill Tribes of Northern Thailand,* is the son of a well-known jungle man, esteemed in his own right as a hunter of tiger and leopard in upper Burma. He is an American of the third generation to be born in southeast Asia. He was raised among the *Lahu* tribal people of northern Burma and attained the honored distinction of the title "Supreme Hunter" bestowed by the *Lahu* upon an individual who has successfully killed the large and dangerous game of the area—the tiger, leopard, elephant, gaur, boar, and bear. Banting could be thrown in, to boot. Gordon has had outstanding experiences in assisting hill people in removing dangerous animals that were terrorizing their area. Few hunters today can write about the jungled mountains of northern Thailand and Burma with the experience and jungle expertise of Gordon Young. A U.S. Army veteran of Korea who has served in numerous sensitive and highly specialized positions for the U.S. Government in Thailand, Laos and Vietnam, Gordon moved to the quieter environs of central California after the war in Indo-China ended.

In a league by himself, as the era of man-eaters and man-killers starts to ebb, is Patrick James Byrne of Calcutta, with whom I had the opportunity of making my first tiger hunt. Pat is the son of a highly decorated Irishman who served in the Scottish Black Watch Highlanders during five colonial wars in the days of the British Raj in India. His mother stems from an illustrious Bengali Brahmin family. His grand-uncle was Governor of West Bengal during the post independence years in the 1950's. Few people still living today have had the experiences in dealing with man-eaters and man-killing rogues that Pat does. His hunting experiences occurred during the height of sports hunting in India in the time between World War II and the closure of hunting in India in 1972. The magnificence of that period will never again be repeated. I have selected a few of Pat's numerous experiences with man-eating tiger, leopard, rogue elephant, and killer boar which describe in detail the horror and devastation which can occur in small rural villages where the people possess no weapons and are largely defenseless in trying to rid themselves of a killer. I am grateful to the *Probe of India* for granting me permission to use excerpts from some of these stories for this book. The editor of *Gamecoin* and *Hunters Quest*, the respective publications of Game Conservation International and the Dallas and Houston

Safari Clubs, have also kindly allowed me to reprint stories which appeared in their publications. Mr. Thomas F. Martin of Bangalore, India, has generously shared his experiences with the Kursela-Kataria man-eating mugger.

As a final contribution, I have added a chapter on the ultimate predator and man-killer—man himself! Just as with man-eating and man-killing wild animals, the practice of man-hunting has been conducted in many parts of the world, particularly Asia, for the purpose of collecting a trophy or, for the more gruesome prospect, of eating the victim in cannibalistic rituals. This also did not end until very recently and may quite possibly still be going on. It is only when man himself experiences what it feels like to be hunted by another man possessing the same faculties and intelligence as himself that he can truly say that he knows what hunting is all about. Only then will he have experienced hunting from both ends of the spectrum.

JHB
J hanging B Ranch
Alamosa, Colorado
July 1988

THE ROGUE OF VILLAGE SIXTEEN

by

Capt. John H. Brandt

A tuskless bull elephant goes on a rampage destroying crops and gardens. Wounded by a spear trap, the bull becomes doubly dangerous and attacks a police vehicle.....

Tahan and his wife cowered in the corner of their little stilted bamboo shack listening to the tearing of branches, breaking of fences and rumbling roars coming from the outside darkness. Tahan was an ex-soldier from the Royal Thai Army and he now felt powerless that he could do nothing to protect his family or his garden from the destruction.

He had no weapon. As the night progressed, the noise gradually diminished into the distance and with the first hint of light Tahan surveyed the shambles. The destruction of his crops *meant that he would have to rely on other villagers for assistance until he could get another garden started. Yet, many of his neighbors had experienced the same misfortune. The elephant bull had been visiting villages in the area for several weeks and every effort to frighten it away with noise, cloth streamers and garden sentries had all been unsuccessful. Tahan quickly decided what he must do. He got on his bicycle for the long ride to the provincial capital at Satun to see if he could enlist help. Something needed to be done—and soon! At least the elephant had not killed anyone so far, but Tahan felt this would likely happen before long.*

The huge, rogue elephant that terrorized villagers in Satun a few days before he was shot. This rare photo was taken by a village school teacher with a box camera.

Satun is a small village in the very southwestern tip of Thailand on the Malayan Peninsula. Across the border lay the Malayan Sultanate of Perlis. Although the province belonged to Thailand, many of the villagers were ethnic Malays. The last remaining Semang pygmies, that I had contacted and described in 1963, lived in the huge forests which extended for hundreds of unbroken miles to the east, north and south on into Kedah, Malaya. In the west, the area was bounded by the Andaman Sea. Satun was the provincial seat of government and it was here that Tahan was headed to see what help the authorities might offer him.

Several days later, at my home in Songkhla on the opposite side of the Malayan Peninsula from Satun, a postman rang at the entry gate. One of the servants let him in. Barefooted, but resplendent in his uniform, he pulled his bicycle up to me and saluted as he handed me the official-looking telegram. He watched me intently as I read it because he had surely read it several times himself and was most curious as to how I would respond to such a ridiculous request as the message contained. He was sure that no one in his right mind would voluntarily risk his life if it was not necessary. He actually considered the request unnecessary. After all, Songkhla was a long way from Satun. The message I had received was from the Governor of Satun Province and said simply, "Elephant problems—Stop—Can you come— Stop—Permission to kill authorized." My postman was disappointed that I made no immediate reply, but, quite frankly, at that moment I wasn't sure myself what the reply should be.

I had no details as to what the request meant or involved, but I had laid the groundwork for this myself never realizing my hopes would bear fruit. In my travels throughout the Peninsula I had often asked local governors if permission could, under any circumstances, ever be given to take an elephant. Since elephant had been considered crown property since passage of a Protection

Act in 1921, no such permits were generally issued. At that time Thailand had no hunting laws and other big game such as gaur, banting, tiger, leopard and boar were taken indiscriminately. Elephant was an exception, however, and, although they were quite common throughout the area, they were rarely molested and few poachers had adequate weapons to tackle something the size of an elephant. As a result, the elephant had little fear of man and, in contrast to African elephant who learned of man as a dangerous predator, the Asian elephant was often extremely aggressive. On more than one occasion I have had elephant intentionally follow my tracks to eliminate me from their environs. The sound of elephant in the vicinity at night in a jungle camp often resulted in many sleepless hours because the malicious intent of the animals was something that had to be constantly reckoned with.

I took my telegram with me to see my friend the *Balat Changwat* (Deputy Governor), who was a dedicated hunter, and asked if he might like to accompany me. Until then, I had only killed one elephant in Africa, but the fact that I hunted, ran around in the jungle and apparently enjoyed such a peculiar pastime had earned me the sobriquet, *Nai Phaa,* or "Jungle Man" in Thai. The *Balat Changwat* was highly enthused but had already made other commitments and regretted that he could not get away. He did insist that I take his .450 double along, which was a big boost to my confidence. I did most of my hunting with a .300 Mag which I didn't at all feel was adequate for the venture before me. By this time I had made up my mind to go and, bidding the Deputy farewell, I went home to get my gear together for the drive to Satun.

The following morning I headed southwest on the road to the border crossing at Sadao and then cut west on the washboard laterite track that skirts the huge border forest called the *don chüok chang*, which refers to the foot hobbles placed on a captive elephant. It took all day to get to Satun because the monsoon rains had soaked the road and washed out several places. There is no such thing as a "dry season" on the Malayan Peninsula and weather variations depend only on which direction the seasonal monsoon happens to be blowing from. As a consequence, the jungles are dense, green and lush. Arriving at Satun, I pulled up to the shop of an old friend, Haji Hassan, who carried the honorific title, *Haji*, because he had made the holy pilgrimage to Mecca. On arriving I had noticed many banners, flags and crowds of people and, on inquiry, Hassan informed me that the Provincial Governor had suddenly taken ill and had just died. One of his last official acts most surely must have been his telegram to me about the elephant problem.

Hassan was well aware of the situation, as he was of everything else that happened in Satun, and said he would make the contacts to arrange the official authorization to kill the elephant and have someone guide me to the villages where most of the damage had occurred.

Early the next morning we left for a drive of some three hours on a fairly good, sandy track passing large rubber tree plantations. The tappers were returning from their early morning chores, and little children enroute to a clapboard school some miles down the road stood at attention and then bowed as we went by, assuming that anyone in a vehicle must be some important government official. Such visits to their villages were a rare event indeed!

We pulled up in front of a small thatched house in a village, and the local *gamnan*, who was chief of several villages, came to greet us and assure us of his gratitude that the elephant was about to be killed. I was not quite as optimistic as he! He piled into the back of the vehicle and directed us to areas, where no vehicle should ever have gone, to show us the damage the elephant had done. The wreckage was awesome and what the elephant hadn't eaten, he had trampled, knocked down or otherwise rendered useless. The garden we examined had been destroyed about ten

Capt. Brandt with the Haji who arranged the local contacts with villagers whose gardens and crops had been destroyed by the elephant.

The Governor of Satun had just died when Capt. Brandt arrived. Here Buddhist priests emerge from the funerary pyre.

A southern Thai delicacy called "Yeh" which is made of sun-dried lizards. When eating them, one must be very cautious of the sharp claws.

days earlier. I asked the *gamnan* where the latest problem had occurred because the elephant obviously had no reason to return to the gardens we were looking at. The *gamnan* instructed a teenage girl to take us to a little jungle hamlet which bore no name but was known as *Muban Sib Hok* (Village Number 16) where the elephant had recently been. It was here that Tahan lived. Again the signs were all several days old and the elephant had not been back since the last rampage. However, Tahan informed me that a frightened villager from two or three kilometers further up the track had informed him the night before that the elephant was seen in the area and had torn up several small sugar cane fields. By now it was late in the afternoon and I returned to the *gamnan's* house where he generously allowed me to spend the night with him and his family. His wife had gone to great pains to prepare a special dinner for me of which the main entrée was a dried lizard called *yeh* and another delicacy known as *kai mut daeng*. This is an

The little girl that showed the author the destruction to her family's garden. On her shoulder is a pet leaf monkey called a lutong.

omelet made from duck eggs and the larvae of an ant that makes huge globular wasp-like nests in the jungle. The sour taste from the ant's formic acid made an interesting seasoning once the proper mental adjustment to what the contents of the dinner entailed had been made.

The next morning Tahan and the teenage girl met me to take me further into the forest. After a walk of an hour or two we reached the trampled sugar cane field. A light rain was falling. As I walked about looking at the elephant's droppings and deep tracks in the mud, I saw something I hadn't expected! On the trail were numerous splashes of blood, and, after more careful examination, I saw additional signs of blood smeared on leaves at the level of my head. I called the men together and asked what they knew of this. All shook their heads and said they had no idea where the blood had come from. I thought, most likely, that one of the

The author talks with badly frightened villagers who had suffered from the destruction of their crops by the rogue elephant.

villagers had shot at the elephant with a homemade muzzleloader, loaded quite likely with chopped nails, rocks or whatever else fitted in the barrel in the absence of a lead ball. After more denials I threatened to leave, since a wounded elephant created a completely different scenario than what I had expected up until then. This brought an instant reaction and one of the men offered to show me what had happened. Together, followed by probably every inhabitant of the hamlet who had never before seen a white man, we walked to the edge of the field where the trees bordered the clearing.

A trail led into the jungle, and the villagers said that for three nights the elephant had emerged from the forest on exactly that same path each time. In a desperate effort to rid themselves of the elephant they had created a diabolical trap with a trip wire across the path. This, when released, triggered a thick bamboo spear which was driven forward about 5-6 feet from the ground. The trap had been set off and the ground showed considerable blood which continued on into the jungle where the elephant had run to escape after the bamboo missile had been driven into its body. We backtracked and shortly one of the men called to say he had found the six-foot spear with at least one to two feet of the tip smeared with blood. The elephant had indeed been severely wounded, but I doubted that such a wound would kill it unless eventual septic infection would do it in. I had no doubt, however, that we now had a very unhappy and much more dangerous animal on our hands!

I returned to the *gamnan's* house and had just started a cup of tea wondering what the best plan would be when I heard outside a commotion of loud shouting and agitated talking. I looked out to see four heavily armed border policemen excitedly talking with the *gamnan*.

The senior officer was a lieutenant named Amporn who was obviously very, very upset. I took him aside and listened to a most fascinating tale of the misfortune that had befallen him only hours before while I was in the sugar cane patch. The lieutenant and a squad of his men had been on an extended patrol in the jungle chasing illusive remnants of the MPLA (Malayan People's Liberation Army), locally referred to as C.T.'s (Communist Terrorists), who made any travel in this part of Thailand occasionally a dangerous undertaking. The squad had come out of the jungle several kilometers away and had loaded into their vehicles for the return trip to their headquarters. Amporn told me that they had barely entered the rubber plantation area when, without any warning, a huge elephant, trumpeting loudly, had charged out of the dense bushes alongside the track and had hit the vehicle broadside with its forehead. The driver swerved off the road and whether the vehicle tipped over in the ditch, or whether the elephant completed the somersault could not be ascertained, but the vehicle lay on its back. The badly frightened squad of policemen had all managed to jump clear, unhurt, and had broken Olympic track records in heading for cover. No one had fired a shot in the confusion, but they had seen the bull close-up and could vouch for the fact that it was covered on its chest with blood that had formed dark streaks down its forelegs. The elephant was tuskless. This is a genetic peculiarity often found in Asian elephant where some bulls never grow the characteristic long tusks of a mature male. Bulls without tusks often are compensated for this by increased body size.

The lieutenant was happy that no one had been injured in the attack, but bemoaned the paperwork that now awaited him on having to explain how his vehicle had been damaged. It would have been easier to justify an ambush by the C.T.'s than to try to explain an unprovoked attack by a mad bull elephant!

In any case, Amporn now made a profound statement that I hadn't expected and could have done well without. I wasn't given a choice, however, and the lieutenant proceeded to rationalize that since the elephant had directly attacked him he was duty bound to settle the score and, also, since he was a police officer, it was his duty to stay with me to see that no harm came to me. I appreciated his sense of dedication but didn't realize that matters would get worse until he filled in the detail that his entire squad would help me kill the elephant! It was futile to try to explain that vast differences existed between hunting wild animals and setting up a jungle ambush for terrorists. In the latter, his men had vast experience. Since argument was futile, I considered myself fortunate that at least the men were very "jungle wise" and knew how to shoot.

We spent the next day contemplating where our ambush for the elephant might be set up. It was unlikely that the elephant would return to the place where the bamboo spear trap had wounded it. While considering possible places, a man leading a huge, vicious-looking macaque monkey, trained to climb palms to retrieve coconuts for its master, approached us saying that the elephant had been to his garden the preceding night and that the elephant had been there before, always coming down a well-defined jungle track. He offered to take Lt. Amporn and me to the place. We set off in mid-afternoon and arrived at the location several hours before sundown.

We quickly decided to try to place *machans* in trees near the mouth of the trail. Cutting some bamboos we tied some makeshift platforms 30-40 feet up in the trees. It was not necessary to put up much camouflage since it would be dark and highly unlikely that the elephant could see us. Since human scent lingered around most of the gardens, and the track the elephant utilized was also used by the villagers collecting food in the jungle, it would not cause the elephant to be suspicious of human scent. At five o'clock Lt. Amporn and I climbed to our platform and we watched as six patrol officers climbed with their automatic weapons to the other three locations. We gave each other a wave to signal that all was ready and then settled in for the long wait. We had no idea whether the elephant would follow his usual pattern or if he had left the area entirely. The drone of mosquitos kept us alert and frustrated that we could not retaliate with an occasional slap. Sit and suffer was the name of the game.

We were both suddenly startled by a loud swishing sound somewhere behind us that grew louder. Then directly over our heads two huge hornbills flew by in perfect unison enroute to some remote forest roost. As they flew by, the air brushing through their stiff wing pinions grew progressively less and was replaced by sounds of doves and other birds preparing for the jungle night. Soon it was dark, but as our eyes became accustomed to the gloom many features of the trail were surprisingly visible and, as the moon rose around eight o'clock, we could see quite well.

We had been on our *machan* about five hours when we all began to hear distinctive sounds of breaking branches coming from the forest. Knowing how quietly elephant can move, and how they can almost mysteriously appear and disappear in front of your eyes, made it obvious that this particular bull had no inhibition of announcing his presence to the world. He had nothing to fear—and he knew it! The sounds became louder and clearer. Whipping or slapping sounds seemed to be made by the bull carrying a branch in his trunk which he slapped from side to side on the track or by the flapping of his ears striking against his shoulders. Amporn nudged me as he and I both simultaneously became aware of a huge, bulky shadow coming down the path. We could see the elephant when he was about thirty yards away. The bull approached at a steady walk totally indifferent to any precaution that animals might normally take. His approach was so rapid that we quickly realized that unless we planned to shoot quickly, the bull would soon be directly under us or already past us. I glanced at the patrolmen and saw that they were watching us with their weapons at the ready. All four *machans* were within a radius of 15-20 yards of where the elephant was now walking.

I shouldered the .450 and aimed as best I could in the darkness at a point directly in the middle of the shoulders hoping that I could break the spine, or that at least the bullet would penetrate the lungs and produce a fairly rapid kill. In this instant, before I pulled the trigger, I recall hoping that the police would not open fire until I had shot, but the men had been well-trained and everyone patiently awaited the propitious moment. As the double rifle went off, it was only a split second before the automatic rifles cut loose. Every terrorist within hearing distance must have had instant heart failure! The rifles with jacketed military ammo must have penetrated well and, in the noise, the sound of bullets impacting on flesh may have been more imagined than real. In a flash it was over! Probably close to a hundred rounds had been fired and the elephant was gone. We listened intently but could only hear the soft rustle of leaves falling from the trees where the bullets had clipped them loose. We felt certain the elephant did not survive the barrage as we heard no further sounds.

Shortly, we saw lanterns approaching from the hamlet and as the lights came closer we shouted a warning to the villagers to approach with extreme caution since we didn't know what had happened to the elephant. Carefully, we climbed down and, rather than run the extreme risk of following the elephant in the darkness, we returned to the village about midnight to spend the rest of the night wide awake, drinking tea, waiting for sunup.

As soon as we could see, we took off on the trail until we reached our *machan*. A fairly steady rain had begun to fall and we were soon all soaking wet. In the early light we could see huge splashes of blood and carefully followed the spoor into the gloomy forest.

We had gone about 150 yards when suddenly one of the patrolmen gave a startled shout which made all of us jump with guns at the ready. The patrolman had deviated slightly from the track and was about to climb on what he thought was a huge gray boulder for a better view, when he realized that the boulder was an elephant.

The bull had gone a few yards further on the trail we were on and then had circled back about 15 yards when death had caught up with him. He had rolled into a small gully and was lying on his side with several jungle vines and vegetation festooning his body. From only a few yards away he was totally invisible and the policeman's mistaking him for a rock was highly understandable. We had heard no noise of his falling because of the cushioning vegetation and the fact that sound does not travel very far in dense jungle.

The lieutenant and I congratulated one another on our success. The villagers were ecstatic. Amporn fingered a small metal Buddha amulet around his neck called *luang pawthuat*, which most military and police carry since it protects them from bullet and knife wounds. We both laughed as we wondered whether it also protected someone from a rampaging elephant. After all, he had been unhurt in the attack on the vehicle; but he did admit that he hadn't been anxious to test it a second time with the bull who now lay dead at our feet.

THE DASINGABADI ROGUE

by

Pat James Byrne

as told to

Capt. John H. Brandt

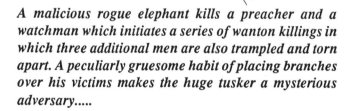

A malicious rogue elephant kills a preacher and a watchman which initiates a series of wanton killings in which three additional men are also trampled and torn apart. A peculiarly gruesome habit of placing branches over his victims makes the huge tusker a mysterious adversary.....

It was early on a Sunday morning in April of 1953. The days were becoming oppressively hot, and Augostino, a lay Catholic preacher, decided to leave early to avoid the heat on his mile walk to the village of Dasingabadi. He was pleased with his accomplishments in that two families from the hamlet had already converted to Catholicism. As he walked, he went over in his mind what he would include in the sermon that Sunday morning.

The Pangali Ghat road that led from his house to Dasingabadi wound through some dense forest in a hilly section. The thick overhanging boughs made a shaded arch and gave a pleasant coolness to the road. Although there were tiger and leopard about, there had been no problem for quite some time and Augostino had no fears as he started into the forested area.

Reaching the summit of a small hill he saw some distance ahead of him an elephant feeding on the roadside. Augostino was not alarmed because elephant were often encountered and moved away quickly after sensing that a man was approaching. As he watched he felt a cool breeze hit the nape of his neck which, within seconds, carried his scent to the elephant. Then the unanticipated and abnormal happened. The elephant jerked its head on catching the tart human scent and raising its trunk started ambling up wind on the roadway directly toward Augostino. As Augostino turned to escape, the elephant let out an earsplitting trumpet and broke into a run after the fleeing man. Although an elephant may appear bulky and ungainly, on an open course he can easily outrun a man. It took only a moment for him to close the distance between himself and the desperate, dodging preacher. The elephant raised his trunk and with one horrifying swipe, like a man crushing a fly with a fly-swatter, he knocked Augostino off his feet and sent him sprawling into the roadside ditch. Stunned and unable to move, the man saw the elephant loom above him and then kneel on him before all went mercifully black. The enraged elephant tore the crushed and flattened remains into pieces and then, following a mysterious ritual incomprehensible to man, proceeded to cover the mangled remains with branches and leaves. Still grumbling to himself, the elephant then turned and disappeared into the dense undergrowth.

Ordinarily, this horrible scenario would have been pieced together later from the physical evidence at the scene. However, in this case, there had been a horrified witness who had watched the entire episode from a nearby hillside.

Ananda was a woodcutter from the village of Pangali who had gone to the forest that day to retrieve a few *sal* logs that he had illegally cut a few days earlier. He felt, since it was a Sunday, that most likely the forest guard would be taking a day off and he would be relatively safe from being caught. What he saw that morning was not a part of his plans and, after regaining his wits, he hesitated for a moment wondering what he should do. If he reported the killing he would compromise his reason for having been in the woods. After assuring himself that the elephant was now gone and he could see or hear no further movement in the area, he decided he had no choice but to make his grisly discovery known and let the authorities know what he had witnessed. As quickly as he could run, he headed for Pangali to report the matter to his headman. Quickly realizing the seriousness of the situation and the imminent possibility that someone else might be killed, the headman ordered Ananda to run to Dasingabadi to alert the police officer stationed there.

Listening to his tale, the police officer in charge decided that, since it was now late afternoon, little could be accomplished before daylight and requested that Ananda remain at the station overnight so he could lead a rescue party to the scene of the tragedy the next day.

Late the next morning, as plans were being completed for a group to proceed to the Pangali Ghat road, a trio of highly excited men ran up to the steps of the police station shouting that an elephant had just attacked the Revenue Bungalow at Dasingabadi and had killed the watchman. They had come as quickly as possible to report the incident which had occurred at 10:00 AM that morning. With the news from Ananda of the preacher's death the preceding day, it appeared most likely that the same rogue had continued his murderous escapade and had continued on to wreak mayhem at the bungalow which was not far from where the first killing had occurred.

The group, led by two police officers carrying old, obsolete rifles, entirely inadequate to stop an elephant, moved cautiously and apprehensively toward the area where the woodcutter said he had seen the killing take place.

Entering the darkened forest area, they saw some distance ahead of them a brush pile in the middle of the road. As they came closer it quickly became apparent from the smell of decaying flesh that this was the funerary shroud of branches and leaves which Ananda had seen the elephant place on top of his victim. Kicking the branches aside, they saw beneath it the fly-encrusted, bloated and blackened body parts of Augostino that the elephant had mangled and mashed virtually beyond recognition. The sight sickened the entire group and it was only with considerable determination that the remains were gingerly placed into a sack to be transported back to Dasingabadi.

With great haste the rescue party trotted back to the bungalow where the second reported killing had occurred. Arriving at the bungalow, the two policemen reported in to the senior police inspector who was already on the scene. Describing what they had found, he motioned them to the back of the building where they again saw an identical mound of branches. They did not have to be told that it would also contain the same crushed fragments of a human being, much as they had found on the roadway. Assisted by some village men, the watchman's corpse, or what remained of it, was also removed and turned over to relatives for cremation.

While they were completing their investigation and talking with the villagers, a man bicycled up to the bungalow with a report that another killing had occurred the previous night at a place about ten miles from the Bungalow. The police learned that a man from the village of Sakerbadi had not returned home the previous evening. Not knowing about any of the other killings, the villagers had waited until daylight and then had walked up the track that they felt the man would have used on his return home. They shouted his name loudly but received no reply. Soon they too found a pile of branches on the trail permeated by the same sickeningly sweet odor of decomposing flesh. The buzzing of the flies around the brush pile left no doubt that they had found the body of the missing man. Just then above them on the hillside they heard the loud trumpeting call of an elephant, and in one movement the entire group fled the scene leaving the body where they had found it. The headman had ordered one man with a bicycle to go to Dasingabadi, by a long circuitous route, to advise the authorities of the killing.

The police were now extremely concerned. Three deaths in a twenty-four-hour period, presumably all the murderous work of one elephant, prompted them to contact the District Officer (Collector) to request his authorization to declare the animal a rogue, which, as a public nuisance and danger, should be destroyed immediately. The D.O. wasted no time in issuing the required proclamation and invited interested hunters to come and undertake the task of killing the animal.

Almost a month had passed since the killings had occurred before I heard of the incidents. A telegram had been sent to my home in Calcutta by the headman of Nuagam requesting my assistance if I could get away. He said others had attempted to kill the rogue but all had been unsuccessful. I made plans to go look at the situation, although at that time even getting to Dasingabadi was a major undertaking. The village lay almost forty miles from Nuagam, which is where any road worthy of the name ended. Beyond Nuagam the only means of transport was by walking or, if available, by bicycle. On my way, I contacted the District Officer who issued me a permit to shoot the rogue and also kindly gave me a letter of introduction to all police units in the area asking for their assistance and cooperation, which proved to be most helpful to me.

With the aid of the headman I hired six men at Nuagam to help me transport my camp gear and equipment. I bicycled but often thought that perhaps walking would have been much wiser. It took several days to reach Dasingabadi where we arrived on the morning of May 26th.

The villagers were elated at my arrival and quickly told me that the rogue was still in the area, and that only three weeks earlier he had attacked a caravan of bullock carts hauling logs. He had smashed two carts, killing the bullocks who were unable to escape his attack. Although the cart drivers had escaped injury, an innocent villager, who just happened to be in the right place at the wrong time, had not been as fortunate. The enraged elephant had caught him in his trunk and smashed his body into a pulp against the base of a tree.

The police at Dasingabadi confirmed the event, and they and the villagers helped me bring my gear to the Revenue Bungalow where I would stay. It was a bit disconcerting knowing that the rogue I was after had attacked the very building where I was to sleep and had killed the watchman there only a few weeks earlier on the first day of his rampage. One of the men who was assisting with the gear broke away from the group and came over to me. He was small and wiry and had a huge grin on his face as he saluted me. Speaking very quietly, as a jungle man should, he introduced himself as Bana. He added that he was the village Shikari and would be my guide and assistant in hunting the elephant. Glancing over his shoulder at the police officers, he added that he was also the village poacher and knew the jungle well. With great confidence he said that he had no fear of elephants and, with himself and I as a team, we could soon slay the rogue. I liked his demeanor and sincerely hoped he was right.

Since the attack on the log carts almost a month earlier, there had been no reports of the elephant, and no other known killings had occurred. As the days went by without any reports, it was difficult not to entertain thoughts that perhaps the elephant had moved away, or died, or changed his habits. All were unlikely, but the frustrating inactivity of waiting made me wonder if I had made a mistake in setting up my headquarters in Dasingabadi. Hunting of rogues is always an exercise in patience, and I did my best to familiarize myself with the area and to keep occupied while awaiting news of another attack by the monster.

Bana and I examined numerous old spoor, presumably of the rogue, in the jungles surrounding Dasingabadi. None were fresh but, from the easily visible tracks in the soft dirt near where the bungalow guard had been killed, we estimated the size of the bull at about ten feet tall. The rogue was a huge animal and one of obviously enormous strength. Unfortunately, diligent searching, day after day, produced no fresh tracks or signs of the elephant in the vicinity. Another week went by and my time was running out. I had almost decided that unless some report came in within another day or two I would start the return trip to Nuagam. Then the next day, what I had been awaiting happened!

A messenger brought a letter to the bungalow from the police officer stationed at a village called Bamnigam, which was located almost twenty-four miles from Dasingabadi. The letter contained a detailed description of a killing by an elephant in the area a few days earlier. The fact that the killer had placed a mound of boughs over his victim confirmed that it must be the same rogue who obviously covered great distances in his travels. I debated on what the best course of action might be since it would take me a day or two, at best, to reach Bamnigam. Bana and Boliar Singh, the headman, discussed the situation with me and said that from their past experience the rogue rotated at fairly frequent intervals throughout his range. Since he had been away from Dasingabadi for several weeks, it might be possible that we would be on his circuit, and he would likely soon return. I knew they were also concerned about my leaving which meant they would be left to face the rogue alone without weapons. We agreed to stay another few days to see what happened since the news of the killing from Bamnigam was now already several days old. It proved to be a wise decision. It was now June 3rd.

The day was one which would long be remembered by me, as well as all the local villagers, because an enormous storm crossed the area that night causing extreme damage to homes and trees. For a time I wondered if the bungalow would survive the lashing rain and cyclone-force winds. About midnight the storm abated somewhat, and when morning arrived I could see trees uprooted all around the compound. A large mango tree had blown over on the roadway, and a number of village women were already there with baskets retrieving the mangos which now were suddenly so conveniently within reach. There was much talking and laughing outside from all the people when suddenly a loud shriek of fear from the women quickly brought me to the verandah to see what had happened. Wanting to be prepared for any emergency, I automatically grabbed the .470 Express rifle propped next to my bed. Stepping onto the verandah I was thankful for this intuitive action because on the roadway was the very creature I had spent so many weeks waiting for. It was as if destiny had arranged this introduction with the rogue on my doorstep! I motioned Bana to remain indoors as I stepped into the yard.

The bull, a huge tusker, was still some three hundred yards away, and, since the wind was blowing towards me, he had not yet scented or seen me. He was also still somewhat distracted by all the scurrying villagers who had run at the first indication of his presence. He had made no attempt to catch anyone, and, after moving a few yards toward the bungalow, he veered to the left and disappeared into the brush. There was now total silence. Every villager had disappeared, presumably to the questionable safety of their frail huts.

I was now alone and proceeded one step at a time toward the area where I had last seen the elephant. I tested the wind constantly hoping it would not change direction and alert the rogue to my presence before I had seen him. It took me several moments to proceed one hundred yards but I could make out no sound or sight of the bull.

Suddenly there was a loud crash to my rear, and I swirled, fully convinced that I had passed the bull and that he was making an attack from ambush behind me. What I saw was hardly what I expected! The bull had silently walked through the underbrush and, apparently unaware of me, had entered the compound. The crash I heard was the spectacle of the bull demolishing the roof of the two-room bungalow. He had grabbed a roof timber and with a shaking tug had torn loose the entire end of the roof. Although he could have now easily seen me, he was so engrossed with his demolition project that he totally ignored my approach. I had closed the distance between us to about twenty-five yards when the bull suddenly stopped and became deathly silent. He raised his trunk and picking up my scent, swirled in a lightning-like turn, and thundered across the small compound toward me. His charge came as no surprise. I had expected it and was prepared.

Before the bull had taken three steps, the heavy bullet from the left barrel slammed into his forehead stopping him in mid-stride. He sank to his knees with a jolt as I fired the second barrel. With a shudder he then rolled quietly onto his side and lay still.

I carefully approached the great tusker, but there was no question that he was dead. I sat on his head catching my breath over the excitement of the past few minutes waiting for my pulse rate to return to normal.

My cook and Bana soon appeared from inside the partially-wrecked bungalow and joined me. Within moments, calls of

Dasingabadi villagers crowd around the dead rogue tusker with Byrne shortly after its death.

curious inquiry came from the jungle in several directions asking, "Sahib, is he dead? Is it safe to come out?" Assuring them that the rogue was dead, we were soon surrounded by a deliriously happy group of villagers. Some men brought a chair in which they insisted I be seated so they could put me on their shoulders to dance in triumph with me through the village. I made it on my precarious perch through the festivity which lasted until late into the night. The Dasingabadi rogue was dead! Although neither I nor the villagers would ever determine what had prompted it to take on such murderous and abnormal pathological behavior, no one really cared.

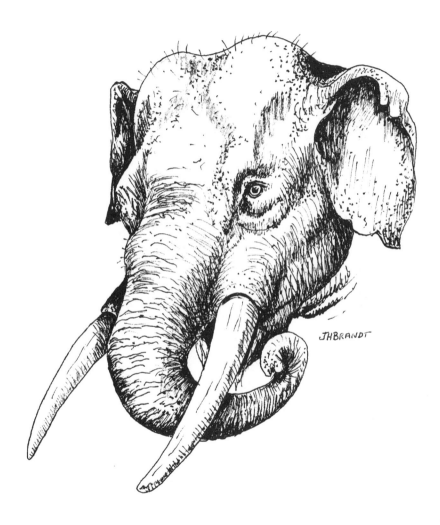

THE THEPAA TIGRESS

by

Capt. John H. Brandt

A new jungle settlement area is invaded by a tigress who becomes overly familiar and aggressive toward people in a scenario in which the ultimate outcome can only have tragic consequences.....

The southern provinces of Thailand along the Malayan frontier are a heavily jungled rain forest with high mountains that gradually dip to the narrow coastal plain bordering the Gulf of Siam. Most of the people live along the coast where agriculture can be combined with fishing. The forests are sparsely inhabited and visited only by occasional bands of Semang pygmies wandering in from Kedah in Malaya or Naratiwat Province of Thailand. Wildlife was abundant including herds of elephant, tiger, leopard, sambar, boar, and even a rare rhino or two. Most dangerous of all of the forest dwellers, in the decade or two following World War II, were remnants of the Malayan Peoples Liberation Army (MPLA) who had waged an unsuccessful war against the Commonwealth to make Malaya a communist state and had been driven by the military into the remote border forests of south Thailand. They were a constant hazard for anyone traveling the area and often made jungle excursions an exercise in survival.

The author with some Semang pygmies who lead nomadic lifestyles in the deep mountainous rain forests of south Thailand and Malaya.

South Thailand's border provinces were an area suffering from governmental neglect because the people spoke Malay,

Brandt and his mahout could reach many places in Thepaa only by elephant.

wore turbans and sarongs, didn't eat pork and were followers of Islam in a predominantly Buddhist country. All this made them a susceptible target for communist insurgents.

Strange concepts about the forests, bordering on superstition, existed including a belief in jungle dwarfs with no knee caps who could only regain an upright position after falling by pulling themselves up on trees. Or a profound belief in the existence of a highly venomous giant black cobra that was unrecognized by science. North of my home in the coastal village of Songkhla, everyone was convinced of the existence of dwarf elephants, only as big as a horse, that were supposed to live in the tangles north of the inland sea of Thale Luang. More profound was a firm belief in people who could turn themselves at will into a were-tiger; although often, in such instances, the tiger's stripes ran horizontally rather than vertically.

The Thepaa settlement was freshly hewn from the jungle to resettle land-hungry farmers.

In this area the government had initiated a policy of chopping down and burning the verdant forest in the misguided belief that once the forest was removed this would then result in productive agricultural land where Thailand's surplus populations could be resettled. That this destruction would be counter productive fell on deaf ears as it has ever since throughout the rest of the developing world where the destruction of the rain forest has continued unabated to this day, from the Amazon and Africa to Indonesia. The government's plan was to resettle land-hungry city dwellers on such tracts where they would plant rice, mango, bananas, papayas and start rubber trees. Such settlements were called *nikhoms*.

The *nikhom* of Thepaa was on the main dirt track between Songkhla and the primarily Malayan Province of Pattani on the east coast. It took its name from the Thepaa River which ran due north from its headwaters on the Malayan border at an elevation of almost 2,000 feet. Parallel to it on the west was the Sakhon River. Both emptied into the Gulf of Siam at Ban Pak Chong.

The *nikhom* lay between these two rivers. The constant burning of the forest gave the atmosphere an eerie gray shroud that at times virtually obstructed the sun as it hung like an orange orb in the sky.

For a number of years I was virtually the only white man, locally referred to as "European," in the area except for a few widely scattered missionaries. It was, consequently, quite an event when in the early 1960's several western countries including America and Great Britain started sending volunteers to help the Thais with various development projects. One such individual landed at Thephaa *nikhom* an agriculture advisor to help the Thais grow crops, although he confided to me that the only bananas he had ever seen were on the shelf in the Safeway store and all his knowledge of chickens was from those distributed by Col. Sanders. Nevertheless, to have someone with whom one could relate and share experiences was a welcome diversion. Hal and I soon became close friends.

On visits to Pattani I had to pass the *nikhom* and would often stop by for a visit. One day, as I pulled off the track I passed a woman sitting by the roadside with a mallet slowly and methodically pounding on a rock to create gravel. The small pile next to her caused me to wonder how long it would take for her to create enough small stones to gravel the road. But many women were doing the same thing and, in timeless Asia, eventually the road would be slowly, very slowly, surfaced. Without heavy equipment or rock crushers there was no reason for the villagers to think it was done in a different manner anywhere else in the world.

Hal met me as I reached the cluster of houses and we were soon joined by a number of village men curious about the *farangs* and their manner of speech, dress and appearance. Particularly fascinating to them was the fact that Hal and I had hair on our arms and legs which they did not possess. Exploratory stroking and some yanking to confirm that the hair was real, quickly followed by a friendly smile to show there was no malicious intent, was a common experience, although it did make one feel somewhat like a chimpanzee in a zoo.

The village headman, the *phuyaiban*, joined us for tea and in the course of the conversation I asked him my favorite hunter's question, "Have you seen many tiger in this area lately?" The responses were often uniform, indicating either that, "no there are no tiger in this area" or "tiger can only be found nowadays in such undeveloped areas like Malaya or Burma, but not here." I often felt that villagers considered it a mark of a barbaric environment to acknowledge that dangerous wild creatures still lived in the area and would consequently point the finger at other areas, any area, but certainly not theirs! Perhaps they were only trying to reassure me, thinking that perhaps I was asking because I feared that a tiger might jump out at me from behind a village hut. I never did figure it out. More likely, the villagers were often totally ignorant of what large carnivores, tiger and black panther, lived around them since such animals can be extremely secretive by nature and are only seen if they want to be seen. Surprisingly, in all of Thailand, tiger are generally more common than leopard but they very rarely turned man-eater. The eating of humans by leopards, in contrast to this often common practice in India, was almost never reported. Both tiger and leopard had an abundance of game, primarily wild pigs, to feed on and very seldom became cattle killers. Quite likely this was due to the fact that they had had minimal associations with humans until the development of the *nikhoms*, and domestic stock was not common. Surely this was about to change since Thepaa was only some 150 miles northwest of Trengannu in Malaya where the famous tiger hunter, Lt. Col. A. Locke of the Malayan Civil Service, had found it necessary to successfully kill several tiger that had taken up the gruesome habit of man-eating only a few years earlier.

Man-eaters, it must be clearly understood, are a different entity than man-killers. The former is an aberration in behavior brought on by age, crippling or a pattern learned by cubs from their parents. The latter, man-killing, is a normal reaction if an unfortunate victim happens upon the tigress with small cubs or stumbles inadvertently into a tiger on its kill. In such cases a spontaneous reaction from the tiger may result in death but the concept of eating its human victim would probably repulse a normal tiger. Notable exceptions might be used in the instances where close daily interaction with humans occurs causing the normally shy tiger to become overly familiar with man. Once its inherent fear is overcome, the situation is an accident waiting to happen. Should such a tiger strike at village livestock and then be confronted with a stick and stones attack by the herd boy it might retaliate and soon realize what a totally defenseless and easily killed creature man actually is. Similarly, should village children or adults working their fields accidentally encounter a hunting tiger, their first fright reaction would be to run which would also easily provoke a spontaneous attack by the tiger. Such a situation was soon to present itself at the Thepaa *nikhom*.

That evening as I started to leave for the drive to Pattani, Hal decided to join me. In Pattani is located the famous Buddhist Temple called Wat Gornio. Annually the Buddhist effigies from the temple are placed on scaffolds and paraded through the village carried on the shoulders of true believers. The effigies command the devout bearers to enter homes and shops and eventually to reach the shores of the Pattani River where the scaffold and a multitude of zealous followers enter the water fully confident that the statues will keep anyone from drowning in the swift current. After the immersion ceremony, the effigies return to the temple where they are carried through beds of red hot coals and roaring flames by barefooted bearers. Others are carried up steps made of sharpened sword blades. If one is pure of soul and appropriately pious, the effigies supposedly prevent burns, cuts or drowning. We intended to witness this event which is unique to that part of the peninsula.

We left the village near dusk. We had hardly driven beyond the cleared area when we both spotted a large black animal sitting in the middle of the road. At first glance we thought it was a dog. I flipped on the headlights and the bright green reflection of the pupils quickly identified the animal as a large black panther. We were no more than 500 yards from the village where we had just been told that tiger and leopard no longer existed there! Black panthers are the predominant color form of leopard on the Malay Peninsula and spotted leopard are exceedingly uncommon. As we pulled closer, the panther made a graceful leap into the bushes on the roadside and disappeared. The animal, quite likely, had long lived in the area adjoining the village but, by hunting only at night

The Buddha effigies are annually walked through the fire in Pattani Province, but the devout are protected from burns.

45

when the villagers were not walking about, it had probably rarely, if ever, been seen by its human neighbors. So much for local wisdom among people who rarely ventured into the forests except to chop down trees and clear the *ladangs* for garden crops.

It was about three weeks later that a bus laden with trussed up pigs, caged chickens, numerous bicycles, and festooned with crowded passengers hanging on the top and sides, pulled up in front of my house on the edge of town. A lone passenger got out and approached my gate which the servants quickly opened. Hal came to the verandah and breathlessly described a tale which began with, "You won't believe what just happened in Thepaa."

Within a few days of Hal's return from Pattani, the village men going to tap rubber trees in the early morning had seen a tiger cross the trail in front of them. The tiger had shown no aggression and had been seen on several more occasions. Within the week, some small children on their way to a nearby village school had also encountered the tiger in the middle of the main track. Everyone had been extremely frightened and when the tiger had let out a roar everyone had turned to run. The tiger had not followed. When one of the braver children looked back in his flight, the tiger had vanished. The tiger had become so visible that the villagers, who usually refrained from even using the word "tiger" in conversation, utilizing instead euphemisms like "the old man," "the striped one" or "Uncle," for fear that saying "tiger" would prompt the appearance of this dreaded animal, were now talking each evening about the tiger which was being seen almost daily. Some villagers were even saying that possibly this tiger was a "were-tiger" and that perhaps some unknown local person had made what is known as *sang phaa* which enabled them to turn themselves from a man into a tiger that roamed the forests at night. Everyone was starting to feel complacent about their new neighbor since no one had been hurt and no stock had been killed. At night the tiger had even serenaded them occasionally from the adjacent hill tops.

Where this tiger had come from or why it was wandering through remains a mystery. It was during the monsoon and the normal mating season was underway. It may have been that the tiger was making a circuit looking for a mate. The nighttime roaring would tend to bear this out. A possibility also existed that a less visible resident tiger might have died a natural death or been

The only way across the jungle streams was on precarious bridges made of rattan and split bamboo that regularly washed away in the monsoons.

killed, unbeknown to the Thepaa villagers in another locality, and this new tiger was moving into the vacant area. The behavior of the tiger was very unusual and aberration spelled danger. The complacency the villagers were beginning to feel was about to change.

One evening during the third week of the tiger's appearance, a village woman counted her pigs which normally wandered about foraging in the forest. One was missing, but no one at the time associated this with the tiger and assumed the errant pig would turn up again the next day. When it didn't return a search was made but nothing could be found. However, some suspicious individuals were starting to consider that maybe this resident tiger was not so innocuous after all. The following evening their suspicions were to be confirmed.

Hal had been visiting a remote part of the settlement south of the main track which was located in an area where even the district headquarters at that time could only be reached by foot or on elephants. His temporary lodging in the village was a small stilted house with split woven bamboo walls and a fragile thatched roof. He had left his bicycle by the small ladder leading to the main floor. Since there was no electricity he had gone to bed when the early jungle darkness had set in. A domestic water buffalo and a white cow grazed in the small clearing in front of his hut.

Within minutes of adjusting the mosquito net, a horrible commotion occurred outside with painful bellows from the cow amidst chilling roars from what could only be a tiger! With no weapon, Hal had cautiously looked out and could make out the death struggles of the white cow, which was still visible in the early darkness as the tiger pulled it down, directly next to his bicycle. Within a few moments the movements ceased and then the tiger made an attack on the buffalo which was tethered and had stupidly witnessed the attack. Tiger will rarely kill more than one animal for its victim, but while the killing lust is running high the second victim may be attacked and killed only because it didn't have sense enough to get out of the tiger's way. As the buffalo was also killed, Hal began to logically wonder if he might be next. It was a most unpleasant concept.

Fascinated by what he was witnessing, he watched the tiger take hold of the head and begin to drag away the limp body of the cow. Soon the sounds of dragging were no longer audible and a deathly silence enveloped the area. For a second he began to wonder if he had imagined it all. After several moments a villager called out, "What happened? Is everyone all right?" Soon kerosene lamps were lit and a few courageous men came out to see what had happened. The sight of the dead buffalo with its neck drenched in blood caused everyone to band together to find solace in company. The entire group of settlers sat in a huddle the entire night too frightened to do anything and wondering what would befall them next. Several chastising comments were made about those that had used the word "tiger" openly in their conversations.

The next morning a group of men walked the several kilometers to the *nikhom* headquarters and reported the incident. Hal was invited to go find me and see if he could bring me back to kill the tiger.

Hearing the story, I felt if we left immediately we might still reach Thepaa in time to sit up over the kill, providing we could find it quickly enough. It was the second day since the kill had been made but, judging from the size of the cow Hal had described, I felt sure the tiger would have enough flesh left to return for several feedings if it had not been disturbed. In fact, tiger from the hot, humid rain forest areas are known to enjoy feeding on rank carrion and I was sure by this time that the carcass would indeed be ripe!

Loading up some camp gear, food and my rifle, we took off for the four-hour drive to Thepaa, arriving in the early afternoon. In the settlement the only signs of the kill were a darkened patch of grass and the head of the dead buffalo. The rest of the carcass had been butchered by the villagers and removed from the site. The drag mark of the cow was readily visible and, accompanied by several village men with machete-like *parangs*, we followed the trail.

The jungle was extremely thick and I felt certain that we would come across the kill very quickly. The cow was a good-sized, sturdy, young animal that had easily weighed 600-700 lbs. and yet the tiger had dragged it with apparent ease through gullies, dense bamboo thickets and over fallen logs. It is hard to visualize the struggle the tiger must have gone through to get its prey so far away from the village or the immense strength involved in moving

The Thepaa tigress pulled down a white cow in the village and dragged it for almost a half mile through the jungle before feeding. After the second night very little was left of the victim to entice the killer back for a third meal.

a limp dead weight. Judging from the relatively narrow forepaws, I figured that the tiger was female. Its display of strength and stamina seemed to indicate that it must be an extremely healthy and normal animal.

When we finally found the kill it was lodged in a small gully where it had become somewhat wedged. The tiger had, I am sure, also become tired by this time and had decided it was far enough away from habitation. By my best reckoning we were at least a half mile from the place of the kill! It almost seems that when tiger kill domestic stock they realize they are performing an unnatural act and try to remove themselves as far as possible so as to hide their sense of guilt. However, this is ascribing human traits to an animal, which is naïve. We were soon to find what a strange animal the Thepaa tigress really was.

The kill was only a few yards from a well-traveled footpath that the tappers followed each morning to some rubber trees. We saw numerous pug marks of the tigress indicating that she also used this path. The cow had been neatly eviscerated and the tail had been severed and lay to one side. The tigress had then begun to feed on the buttocks and had eaten fairly heavily from the haunches and the front shoulder. She had apparently returned both nights to feed.

There were no reasonable trees on which to build a *machan* platform close to the kill and I did not relish a ground sit-up for the tigress. This was reaffirmed when the village men pointed to the ground where we were standing. It was literally alive in concentric circles of movement which were closing in on us! The ubiquitous land leeches of the Peninsula were zeroing in on our bodies and we constantly had to shift position as we contemplated what we were going to do next. Land leeches, during the rains, are a scourge that virtually preclude travel in some areas. Small, slimy, bloodsucking creatures that raise their heads searching the air for prey and moving along like inchworms, they are able to enter shoelace holes or attach themselves to any exposed area of skin. The bite is not felt and often the first indication of their presence is a squishing in your boots from the blood draining from feet and legs. Once the leech is removed, the site of the bite often becomes septic and develops into ulceration without proper care. Tiger were not the only creatures to fear in the forest!

My village escorts felt firmly confident that the tigress would return to her kill by the footpath and, since this was the only place we could build a platform in some huge trees, I agreed to give it a try. Cutting bamboo and *rattan*, we quickly put up three platforms about 30 yards apart along the path some 60 yards from the dead cow. If we had miscalculated and the tigress had chosen another approach all our efforts would have been wasted because we were too far away to see the kill properly.

I had brought four additional flashlights and distributed them to two men who had muzzleloaders. Two men climbed to each platform—one to shoot and one to handle the light. We settled in almost an hour before dark and awaited the tigress, assuming she would come after dark and we would shoot by use of flashlights. I was daydreaming, listening to the jungle sounds, when my reverie was shattered by a loud, earsplitting explosion which sounded twice as loud in the dead silence of the forest. I spun around and looked at my companion thinking an accidental shot had been fired. It was still light enough to see quite well and I saw that everyone was looking at the path. Suddenly I too saw what they had seen and was astounded to see the tigress walking nonchalantly along the footpath toward me as if nothing unusual had happened! The shot from the first muzzleloader had apparently not hit her and she had never looked up to see the source of the noise. As she approached the second platform I became frustrated that I didn't have a clear shot and wondered what would happen when the second gun went off. Another smoking blast and the tigress jumped into the air and looked around her startled by all this peculiar noise but again made no effort to jump into the bushes or to get away.

There has been much written about hunters in *machans* turning their lights onto a feeding tiger who would stare in a startled fashion momentarily at the light and then return to feeding as if nothing unusual had occurred. More bizarre instances have been known of a hunter shooting a tiger over a kill and then shortly thereafter turning on his light to find another tiger feeding on the same kill, also to be shot in its turn. One hunter that I know of had killed three tiger in one night this way which was quite likely a female accompanied by fairly grown cubs. Often animals associate gunshots with normal phenomena such as thunderclaps, rock slides or branches falling from trees. I have had later experiences with both caribou and sheep that totally ignored the sound of a rifle blast.

The Thephaa tigress had exhibited so many peculiar behavioral traits that her reaction to the gunshots should have been no surprise, but I watched in fascination as she continued her walk directly toward my tree, only occasionally looking about somewhat more alert as if wondering what the sharp noises had been all about. Fortunately, it appeared that none of the sundry lethal junk that the muzzleloaders had been charged with had struck the cat and I was thankful that we wouldn't have to deal with a wounded animal. All this had happened within less than a minute and the tigress was now almost under my tree.

I lined up the .300 Mag on her shoulder and fired from a distance of less than fifty feet. A miss seemed impossible but when I fired the tigress made a somersault jump, snarled and disappeared into the brush alongside the pathway. Momentarily, I feared the worst!

Suddenly, one of the men on the other platform, who was more elevated than I, shouted loudly "There it is!" and we turned to see that the tigress had emerged back onto the pathway some fifty yards beyond my tree. It was impossible to make another shot but suddenly the tigress stopped and her head sagged as she slowly sank to the ground. Before I could shout words of caution my audacious village hunters had descended to the ground and were already sprinting past me to see the dead tiger. I sincerely hoped that she was dead, and their enthusiastic shouts soon confirmed that she was indeed down for the count.

We sent a man back to the village on the footpath to get help and soon a group of men arrived with lanterns. Everyone had to examine the tigress to assure themselves that it was indeed an animal and not a were-tiger. When all the stripes ran in the right direction they—and I— were considerably more relaxed.

Word spread quickly that the man-eater had been killed and villagers surrounded the beast to view the dead tigress.

The next morning all the people from several jungle communities nearby came to see the tigress. Many brought their children to see an animal that they had convinced themselves no longer lived in their area. We had to post a careful watch over the dead tiger to keep people from carrying off souvenirs. A tiger tooth, it is believed, will assure the owner of never going hungry; and a claw can be endowed with magic properties to become a *lep mahasanay*, which is an irresistible love magic. A scratch from such a claw arouses instant uncontrollable erotic passion in the person scratched. Yes, a tiger is a powerful creature . . . both alive as well as dead!

The Thepaa tigress is examined by a young villager shortly after her death.

After everyone had been satisfied that the Thepaa tigress was really dead and had begun the return journey to their home, I wished that from then on they would become more observant of the wildlife around them and hopefully become more cautious before so confidently saying, "Oh, no, Tuan, we have no tiger in this area!"

All too soon that prophecy would regrettably become true as the forests continued to be destroyed and the wildlife gradually disappeared from Thepaa.

THE MALISANCHRA MAN-EATER

by

Pat James Byrne

as told to

Capt. John H. Brandt

A tigress, whose long reign of terror has already resulted in over twenty human deaths, becomes brazen enough to attack a public bus, log carts and halts work at a lumber camp. A woman is killed by the man-eater, and her daughter almost brings inadvertent death to both herself and the hunter.....

The Divisional Forest Officer (DFO) of Bhanjnagar, a small town located deep in the forests of Orissa State in a district called Ganjam, looked at me gravely as he handed me my shooting permit to hunt in his area. I knew he had something on his mind, but perhaps knowing I was on holiday caused him to hesitate to burden me with whatever problem was plaguing him. He said,

57

"Pat, do you have a moment that I could talk with you about something serious going on in the area that has everyone half frightened to death? I'm hoping that while you are here you may be able to help us in ridding ourselves of the problem." I anticipated what was about to unfold and sat down with the DFO to listen to his story—parts of which I already knew.

He proceeded to tell me that for several months a tiger had been terrorizing the area and had already killed over twenty people. It had gotten so brazen in its attacks that it had even attempted to pull a victim from an open-sided bus and many people were now even too frightened to go to market on the public transportation. Logging at a nearby lumber operation had come to a virtual halt since the cutters were afraid to go into the forest to work. The DFO went on to tell me that a free permit to kill the tiger would be issued if I would agree to undertake the task while I was there. When I nodded my approval, he smiled with apparent relief and said that it just so happened that he already had the filled-out permit in his pocket, which he passed on to me with a smile and a flourish. My holiday had now taken on a totally new perspective in having to deal with a man-killer!

It had all started one year earlier when I had established a firm friendship with Raju, the Range Officer for Bhanjnagar, who often invited me to come hunt in his area which was well-known for elephant, gaur, sambar, leopard, and numerous large tiger. It was the last year of the war and one delay after another occurred which kept postponing my plans to take a much-needed vacation. In December of 1945, with peace finally at hand and normality returning to India, I contacted Raju and asked if he would initiate the necessary paperwork to allow me to hunt, and make appropriate reservations for me and my servant at the Local Fund Bungalow. The train from Calcutta deposited me at the Berhampore Rail Station from which I caught a small bus for the short ride to Bhanjnagar. At that time the village was still called

Russelkonda by which name it was known until Indian Independence a few years later. Raju had met me at the rail station and on the way to the bungalow had already pointed out to me numerous places where the man-eater had made human kills. Several of the victims had been taken within a very short distance of the village itself. With this information already made known to me, I rather expected that the DFO would ask for my help, which, in any case, I would not have refused. After taking my leave of the DFO, the hunt was to start much sooner than I had anticipated.

Shortly after settling in at the bungalow, I saw a car careening into the driveway with Raju in the passenger seat. Walking quickly toward me, Raju introduced me to the vehicle driver, Gaffur, who was a merchant in the village. The two had been riding along the roadway near milestone forty-seven when they had seen a tiger, which they were certain was the man-eater, standing near the road. They implored me to quickly grab my rifle and accompany them back to the area in hopes of shooting the animal. I crawled into the small Fiat with them, but had few illusions that the tiger would conveniently wait for us to return, and I was certainly not fully convinced that this was the man-eater since there was a sizable population of tiger in the area. The two men did assure me though that milestone forty-seven had a direct relationship to the man-eater who had made kills in the vicinity and had been seen there several times. What predilection he had for that particular spot I was never able to determine, although I later searched the locality carefully. Reaching the place where Raju and Gaffur had seen the tiger, I examined the ditch banks for tracks or other signs but could locate nothing. My two companions judiciously stayed in the vehicle while I wandered about looking for signs. Concluding that the venture was futile, I soon returned to the car for the short ride back to the village.

I had just completed breakfast the following morning when the same car came to halt in front of the bungalow. Gaffur called out to me that a report had been received that the tiger had made a kill just outside the nearby village of Bamsalundi. On the way we picked up Raju and soon pulled up in front of a thatched hut in the village where the kill had occurred. A small, wrinkled man in a tattered *dhoti* and a turban came out to greet us and identified himself as Hari, the village's primary hunter. Other men quickly joined us and told me that early that morning, as they were

A herdsman, out with the village cattle, had just been pulled down by the man-eater.

beginning their daily work, the village cattle had suddenly come running in a terrified herd back to the safety of the corral. A few of the men had run to investigate and were shocked to see a tiger in the grazing fields behind the village standing over the prostrate body of the herdsman. Seeing the men approach, the tiger had grabbed the herdsman, who was still apparently alive, by the small of the back and bounded off to a nearby jungle area several hundred yards away. Unable to do anything, the men had watched the herdsman feebly jerking his arms and legs as the tiger ran off with him. He had made no sounds or cries for help.

I asked the men to take me to the place of the killing and we easily followed the tracks of the tiger through the field which was liberally sprinkled with splashes of blood. I looked at the tracks carefully and decided we were dealing with a tigress. The description of the animal seen by the men confirmed that the killer was a female.

The tracks entered the jungle where tracking became a bit slower. I was proceeding very carefully when a langur began calling in a tree in front of us. I whispered to Hari to take two of the village men to see if they could locate what the langur was signalling his alarm call over. Just as Hari was about to move off we all halted in mid-step because a very audible dragging sound could be heard directly to our left. Suddenly the sound ended in a dull thud as if something had been dropped. Raju and Gaffur quickly decided that tiger hunting was not their calling and were most happy when I suggested they return quickly, but quietly, with the other men to the village. Needing no further encouragement, everyone was instantly gone and I was alone to continue my stalk.

I checked the safety on my 10.75 mm Mauser and carefully made my way toward the sound I had heard a few moments before. The jungle was now totally silent but I felt sure the man-eater was nearby, although I hoped she was not aware that I was on her trail looking for her.

Within a very short distance I came to a small, dry *nullah* which was about four feet deep where the water had cut through in the rainy season. In the bottom I found the body of the herdsman. The tigress had chewed off both legs but had not had time to do any serious eating on the corpse. I knew she was still nearby since the langur continued to call at regular intervals. I searched the thick bush above me as I made my way back up the embankment since I didn't relish the idea of the tigress coming in on me from above as I examined the body. There was a large mango tree with dense foliage almost directly above the body and I climbed up onto a comfortable branch in hopes that the tigress would return to her kill as soon as the jungle quieted down.

I had only been in the tree a few moments when a clear, purring sound came from the dense bush at the left of the tree. The sound could only be from the tigress but, although I strained my eyes searching, I could see nothing. The sounds continued for several moments but the tigress never showed herself. Soon another sound became noticeable in the distance as I began to recognize the noise of approaching voices. Soon the mango tree was surrounded by at least fifty villagers, all talking loudly to bolster their courage. In the group was an Assistant Sub-Inspector of Police and two constables. I climbed down from my perch and Raju and Gaffur introduced me to the officials who were already instructing some village men to make a stretcher on which the body could be placed.

On the way back I told Raju and Gaffur that their untimely arrival had probably deprived me of an excellent chance to shoot the man-eater who had been obviously intent on returning to the body. Everyone apologized but said that once the report had been made of the killing that the police had arrived and had ordered the villagers to take them to the body. Not knowing that I had already found the herdsman and was sitting up over the kill, they had followed the orders of the police. Sitting up over a dead human body is always a touchy situation and I decided that the following day I would attempt to put out some normal bait for the tigress. I procured four buffalo calves which I planned to stake out at suitable places to entice the tigress to make another kill.

While buying the calves, I met a timber contractor who told me that his business was in serious trouble since four of his lumbermen had already been killed by the tigress and that his other men were now refusing to go into the forests. He made a proposal that I go along with him to his felling area since the tigress had chosen the place several times to make a kill. It made good sense that I might have a very good chance to shoot the tigress and at the same time provide some protection for the lumbermen.

Although I spent four days in the area, the man-eater made no attack. However, judging from the frequent alarm calls of monkeys and birds, I felt sure that the tigress was patrolling the area in full daylight but had decided, for reasons known only to her, not to make a kill among the tree cutters. The workers were, of course, quite pleased with that decision on the part of the tigress!

But the tigress, by now in need of a meal, decided on the fifth day to make her attack at the small nearby village of Kontaipalli where she struck down an elderly village woman. The unfortunate victim had been sitting with other women within fifty yards of the village when she had noticed a movement in the shrubs nearby. Foolishly, she got up to investigate; the other women were horrified witnesses as the tigress pounced on the woman and, with virtually no effort, took her in her mouth and disappeared into the bush.

A messenger was sent to inform me in the lumber camp and I quickly made my way to the village. It took only a little while to locate the woman's body which the man-eater had only dragged a short way. Apparently, quite hungry, the tigress had already consumed virtually the entire body of her most recent victim. The woman had not been very large, but all that remained was her left hand, which had been gnawed off near the elbow, and her head. I felt strangely that the tigress, with a full meal in her belly, was quite likely nearby, and I thought the possibility of coming up on her by careful stalking was quite good. Although many tiger hunters have made successful one-man stalks in the dense jungle, I felt that when one man has his eyes following the trail, another pair of eyes is necessary to carefully screen the ground ahead so that we wouldn't inadvertently court disaster by blundering into

The other women were horrified witnesses as the tigress took the woman in her mouth and disappeared into the bush.

the sleeping animal. Tigers might justifiably take unkindly to such a rude awakening! Suggesting that the village men retrieve the remains of the woman and return to the village I asked Raju and Gaffur to accompany me on the stalk, which was not very enthusiastically received. I was quickly coming to the conclusion that my two volunteer assistants left much to be desired as tiger hunters. After listening to their unanimous arguments that a stalk was not a good idea, I gave up and we all returned to the village with the small bundle containing the head and left hand.

With no new reported kills, I spent the next two days examining the area around milestone forty-seven to again try to find what attracted the tigress to the area. Raju accompanied me and several times as we bicycled down the gloomy forest trails I had a gnawing feeling that the tigress watched us as we went by.

Several times we would stop as a sambar belled or a langur called to try to determine what they had seen. If the tigress was nearby she kept herself carefully screened from us and, other than that deep-seated chilling feeling that we were being watched, we saw nothing of the man-eater.

Returning one evening to the village I was bicycling about twenty yards in front of Raju when a large langur bounded up a tree a few yards up the road and started a furious, highly-agitated call while he stared into the bushes next to the road. The area was fairly open, and the thick bush had ended on the sides of the trail which was about to join the highway. I slowed down on the cycle and suddenly saw a movement in the bushes that the langur was staring at. In what must have been one movement I stopped the

The tigress kept herself carefully screened from us, but we had the deep-seated, chilling feeling that we were being watched.

bike, tossed it aside with a resounding crash and simultaneously unslung the rifle from my shoulder. I jumped behind the milestone marker and trained the rifle on the bush where I had spotted the movement. Raju, by now, had come up behind me with the shotgun in his hand and, although he had not seen the movement, he quickly had sized up the situation. All this had happened in a matter of less than a minute or two. As we waited, pondering our next move, we could hear the approach of carts coming toward us on the road. Wondering where our tiger had gone, if the movement had indeed been a tiger, we suddenly heard shouts of *"Bagho! Bagho!"* (Tiger! Tiger!) followed by an agonizing bellow of a wounded bullock. We quickly jumped from our vigil behind the milestone marker and ran as fast as possible toward the highway where the shouting was taking place.

The man-eater attacked a caravan of bullock carts in broad daylight attempting to pull a driver off the wagon.

Reaching the highway we noted about thirty carts pulled up on the roadside. Several of the cart drivers saw us running toward them and approached us, all shouting at the same time that a tiger had just made an attack on them. Apparently, our tigress, on seeing us bicycling toward her, had turned from the trail and had seen the carts approaching on the highway while Raju and I had been searching for her at the milestone.

The man-eater had seized on the opportunity with the intention of yanking the driver from the lead cart. Fortunately for him, the bullock had seen the tiger seconds before the driver did and, in panic, had swerved sharply, throwing the tigress' attack off target. Instead of grabbing the intended victim, she had landed on

Instead of grabbing her intended victim, the tigress landed on the bullock which she mauled while the cart driver ran for his life.

the ox which had been badly mauled. The tigress, thwarted in her plans for an easy meal, had then jumped into the bushes leaving the badly bleeding bullock and frightened driver on the highway.

We retrieved our bicycles and, telling the cart drivers to move on, we stayed for a while longer in the vicinity, but the tigress had by now vanished. It soon became dark and, knowing the danger we would be in, we followed the ox carts down the highway back to town.

My holiday time was passing quickly and all my days were now devoted entirely to the problem of the man-eater. I decided that my next best action would be to attempt staking out the buffalo calves again.

Although all the locations should have been perfect, the tigress again showed total disdain and disinterest in what should have been a normal and easy kill for her. The eating of human flesh had apparently become the overwhelming objective in her hunting, and other animals were of little or no interest to her.

I had two days left on my leave before I would be forced to return to Calcutta, and I was beginning to think all my efforts to rid the area of the man-eater would be futile when word was delivered to me that the tigress had made another kill at the nearby village of Malisanchra.

I learned that a woman and her daughter had been out gathering sticks for firewood near the village when the tigress had jumped upon the older woman carrying her off. A hasty search by the village men had not yet located the body of the victim.

While this was being explained to me, the loudly wailing, hysterical young woman who had witnessed her mother's death fell at my feet and begged me to find her mother. She held tightly to my ankle and I had to extricate myself from her grip with some difficulty.

I asked two village men to accompany me in my search for the body and had only just left the village when the screaming daughter rushed after us and grabbed me by the back of my hunting coat begging to be allowed to accompany us to find her mother. I realized that having someone like her along was the very last thing we needed when looking for a man-eater! I called to some village women to come and take the girl back to her hut. I gave them strict instructions to confine the girl until her hysteria was under control and she had calmed down. I regretted that I had no sedative to give her, but at the moment I had more pressing problems because the tigress was known to be nearby.

We quickly found the axe that the mother had dropped when the tigress had attacked, along with a small bundle of sticks which lay under a small pipal tree. There was a large pool of blood on the ground where the victim had been pulled down.

We easily followed the splayed out pug marks of the tigress as she carried off her victim. Large splashes of blood were easily visible along both sides of the track as apparently the woman was not yet fully dead and was bleeding profusely.

I instructed the two village men to stay behind me since I felt the tigress was close by. In carrying off the woman, the tigress had initially kept to a fairly open trail for the first few hundred yards, but she had then veered off to the right along a narrow game trail. I stopped before entering the thick bush to open my rifle and make sure that it was properly loaded. I slipped off the safety on the Mauser and started forward, very slowly, one step at a time.

Before going very far, I found a string of broken blue beads which the unfortunate woman had apparently been wearing. Carefully putting them into my pocket I continued my search. Reaching a bend in the trail I found the woman's bloodstained blouse. A few yards further I saw the woman's sari hooked onto a thorn bush. Everything was blackened with the quickly drying blood.

I looked ahead but could only see clearly for a few yards. I took a step and at that moment made out a very indistinct swishing sound ahead of me on the narrow trail. Listening carefully, I felt the sounds were being made by a tigress carrying her victim through the bush. Although I couldn't see anything, I assumed the tigress was carrying her victim with the head and feet trailing on either side which made the noises as she proceeded through the vegetation.

Walking very cautiously, searching every bush ahead of me with my rifle at the ready position, I could still hear the sounds but could see nothing. Then suddenly a new sensual alert struck me! A small breeze had brought the very recognizable strong body scent of the tiger to me. There was no question of what it was and I momentarily froze realizing how close the cat must be.

I had proceeded less than twenty additional yards when a twig snapped behind me. That was bad enough in itself, but at the same moment something rushed toward me and fell at my feet grabbing me from behind. Needless to say, at a moment like that it is hard to keep from yelling out loud and jumping straight out of your boots, fully convinced that you have become the tigress' next victim. With my heart beating furiously, I now saw what had grabbed me.

At my feet lay the hysterical daughter beating her head against my rubber soled boots as she tightly gripped my ankles. At that same instant there was a shattering roar directly in front of us followed by soft thudding sounds as the tigress bounded off into the bush. It took little imagination to realize the extreme danger we were both in had the tigress chosen to attack us, and I doubt if I could have done much to save us in that instant.

With the adrenalin pumping, I was furious with the foolish, distraught, young woman who had almost caused both of us to be killed. Unable to free my feet, I rapped her sharply with the rifle butt which caused her to let go. She collapsed in a sobbing heap next to the trail. I ordered her to remain where she was and, stepping over her, continued on the trail. Within fifty yards I found the remains of her mother lying under a waist high bush. It was now clear how close we had been to the tigress!

Looking at the body, I saw where the man-eater had grabbed the woman's throat; four huge gaping puncture wounds were visible. The tigress had already eaten the left leg from the thigh down. Most puzzling was a peculiar slash through the right breast. The woman had been full breasted, and the cut had ripped from the nipple to the base of the breast with the precision of a surgeon's scalpel. Whether this was caused by a claw or a fang, I could not determine.

As I knelt beside the body, I heard a low, purring sound indicating that the tigress was returning. Leaving the macabre scene of the mutilated body, I faced the sounds of the approaching cat and backed very slowly from the kill thinking any moment the tigress would break cover. I had the rifle at my shoulder ready to shoot if I saw any suspicious movement, but the tigress did not show herself and the purring had now stopped. I remained for ten minutes, turning slowly in all directions to avoid an attack from the rear. The tension that builds under such circumstances is about as much as the human body can stand, and it was a decided relief when I heard Gaffur calling in the distance from behind me on the trail.

Walking sideways, watching all about me, I called back to Gaffur who, upon hearing my voice, ran quickly toward me. I then saw that he had a number of villagers with him and in their midst was the still-sobbing girl.

They told me that upon hearing the tigress' roar some minutes earlier and not hearing a shot from me, they were certain that I had been killed. They felt there was no need for silence and they had come forward expecting to find my body at any moment.

No one could give me a valid account of how the girl had escaped them to follow me, and everyone expressed deep embarrassment when I explained how close we had both come to being killed by the man-eater.

I felt that by now the tigress had been thoroughly spooked by all the activity and, although we did sit up over the kill that night, we heard no further sounds to indicate that the tigress was still around. I was quite happy to see dawn approach after the excitement of the previous afternoon. I was most anxious to find a bed, any bed, where I could sleep for a few hours!

As evening approached, Gaffur drove by with Raju in the car and suggested that we do a bit of patrolling along the road to Kontaipalli. Since the tigress had been deprived of a proper meal the day before, it seemed possible that we might encounter her while she was out hunting. With no better ideas in mind and since the last staked-out calves had been undisturbed for several days, I agreed to go along.

It was a cloudy evening and rain appeared imminent as we drove slowly along the road, keeping close watch on the jungle on either side. As the sun went down we scanned the roadsides with the powerful beam of a flashlight. Just as we reached the now familiar milestone forty-seven, I felt a tap on my shoulder. My eyes followed the beam of the light and, standing very complacently near the road, I could make out the body of a tiger watching us from behind a small date palm.

I was carrying a shotgun loaded with L.G. that evening rather than the Mauser, and I carefully raised it to my shoulder sighting in on the animal which, transfixed by the light, had still not moved. She was staring intently at us as I squeezed off the rear trigger. My companions also had shotguns and instantly both of them also fired.

In the beam of the light we saw the tigress jump from cover and leap for the road. She only went a few yards before she collapsed in a heap. We watched for several minutes in the car with the apparently now dead tigress clearly visible to us. Carefully we approached the body and after throwing a few rocks at her decided she was safely dead. It took massive efforts for us to be able to load the huge animal into the back of the small Fiat. With us on top of the carcass there was barely room for Gaffur to drive, but we made it back to the village blowing the horn loudly.

Word quickly spread that the man-eater had been killed and the village people surrounded us in a huge throng. We had to put a guard on the tigress to keep the villagers from plucking out the whiskers which were believed to have great magical properties.

The Malisanchra man-eater was emaciated and thin from a badly crippled leg.

As the crowd's curiosity was slowly satisfied and the people melted away, I was able to more carefully examine the tigress. She had sustained nine L.G. wounds from the shotguns in her head, right foreleg and shoulder. It was clearly apparent that the tigress' left hind leg was considerably thinner than the right one. The atrophy had been caused by a jaw trap into which the tigress had apparently stepped some years earlier. It had damaged the bone and crippled her so that she could no longer hunt normally. She was rather emaciated and we all agreed that this unfortunate trauma was what had caused her to turn into a man-eater.

While skinning her the following morning, we also found a number of porcupine quills imbedded in both her front feet. These also may have been contributing factors in making her the terror of the area where she had claimed so many victims.

The following day my holiday ended and, although I must admit that it had been far from relaxing as I had originally intended, I did find considerable personal satisfaction in knowing that the Malisanchra man-eater would no longer terrorize the area.

THE DEMON LEOPARDS OF KRABI

by

Capt. John H. Brandt

Two leopards, one black and one spotted, terrorize a rural area resulting in several unprovoked maulings. Courting turns from blissful tranquility into murderous mischief.....

It usually took three days, with a bit of luck and not much rain, to make the trip from Bangkok, in Thailand's central rice plains, down the long peninsula to the frontier with Malaya. It was a distance of some 850 road miles which zigzagged across the Isthmus of Khra, where the Malay Peninsula geographically began, across to skirt the southernmost tip of Burma at Victoria Point, or Kawthaung, as it was later called. From there southward, the term "road" is used with tongue in cheek. There were bridgeless rivers where elephants had to haul vehicles from bank to bank which then necessitated a day or two to dry the engines out again before being able to proceed. Axle-deep mud often made four-wheel drives spin helplessly. As if all the complexities of travel caused by mother nature were not enough, bandit gangs called *jon pha,* in Thai, operated throughout the area which truly made such a journey an adventure *par excellence*!

Yet, in spite of the adversities, the area was virtually a solid, unbroken rain forest of unsurpassed beauty. No roads and very few jungle villages blemished the interior forests, and game, tiger, gaur, elephant, boar, a few banting and many leopard inhabited the area. I had often flown over the jungles of the peninsula and marveled at the cauliflower or broccoli-like appearance of the giant forest trees that covered the mountains which had not yet heard a woodsman's axe or the roar of a chain saw. I speculated, quite justifiably, that if my plane were to go down, I would quite likely be the first human to set foot on some of the forest paths below me. Further north, hill tribes of several varieties called the jungle home and further south pygmies and other nomadic tribal people still inhabited the dense forests along the border. However, from my home in Songkhla, north to the Isthmus of Khra, there was little else but forest and animals—except for the infamous road called, jokingly, the "Bangkok to Singapore highway!" Travel on the peninsula was most widely done by rail or by plane on the national TAC (Take A Chance!) Airline whose birdmen flew with the finesse of Alaskan bush pilots. Neither the niceties of train or plane, however, were utilized by me as I started down the peninsula during the last monsoon enroute to Ban Pak Chan located directly on the narrow throat of the Isthmus. I had received word some time earlier of a man-killing tiger in the area and I thought it worthwhile to have a look while passing through the locality to see if it might be possible to give the villagers some assistance. I had brought along my rifle, a .30/06 Winchester, and some rain gear which was virtually useless, but which lent an element of psychological support when venturing into the forests.

All my efforts proved unnecessary, because when I arrived at the district office I was informed that the tiger had been shot and killed and that Buddha had now willed that tranquility must return to the area. The unfortunate tiger, who had developed the unsocial habit of killing people, apparently had been wounded some months earlier, and the district officer informed me that the animal's crippled right front leg had protruded at an awkward right angle to the body. The poor creature had only been able to hobble on three legs and I felt thankful that the tiger had been destroyed for its own sake, as well as the well-being of the local villagers.

The journey on south paralleled the west coast where the Burmese Mergui Islands blend into the Straits of Malacca separating Sumatra from Malaya. The few people eking out an

Huge limestone monoliths rise from the tangled coastal jungles of Krabi on the Malay Peninsula in south Thailand.

77

existence from the forests had planted jute and a few rubber trees, but were mostly dependent upon subsistence gathering of jungle produce. The gastronomic highlight of fruit collecting in the area was *durian*, a foul-smelling basketball-sized fruit covered with murderous thorn-like barbs, but which had a custard-textured interior that drove the olfactory senses to distraction, and tasted like a gourmet's dream.

My destination was Krabi, a small provincial seat of government seated on the edge of the mangrove-studded shore of what would become the Bay of Bengal if one paddled far enough west from the coast. I had intended to get there before everything closed up for the night, but my arrival was delayed by a road closure caused by a crowd of people sending off initiates into the Buddhist Priesthood. The festivity was a major event and, in Thailand, virtually every young man spends some part of his life as a priest. Mothers, particularly, encourage such pious action on the part of their sons because merit will accrue to them in the next life—whatever it may be. With shaven head, tangerine robes and begging bowls, the young novitiates are shown great respect and deference by the people. Occasionally, criminals have been known to disappear into this homogeneous throng where everyone looks alike, dresses alike and acts alike, and all they needed to do was shave their heads, don a saffron sheet and beg for others to feed them. No distinction for acquiring merit was made by first identifying who you were feeding—criminal or commoner.

The peninsula communities, at the time, were suffering a major outbreak of cholera, and as I pulled into Krabi I saw the large number of people at the dispensary desperate for medical help. Surveys had established that most villagers first sought help from village medicine men or "shot doctors" who offered deals like "no cure, no pay," but the rapidly fulminating nature of cholera did not allow for such leisurely niceties and the people were forced to seek help from the less popular government-sponsored health centers.

Pushing through the people seated on the steps, I made my way into the examining room where the physician's assistant worked on a patient surrounded by a dozen village people all offering advice, criticism, and voluntary assistance. The concept of privacy in a doctor-patient relationship does not exist in much of rural Asia! My old friend, Prasong, greeted me as he saw me enter and called me over.

We had known each other for several years and had developed a close rapport. Although Prasong was a health worker by profession, he was a most unusual individual in many other respects. He wore a decoration on his uniform indicating service with the U.N. Command in Korea as part of the Thai contingent. He, above all, was an avid outdoorsman and a very accomplished hunter.

This was a rare characteristic in Thailand where no real tradition of sports hunting, as westerners think of it, ever really existed. Since Thailand had not undergone the dubious benefits of colonization, only an occasional teak cutter, miner, or government official had ever hunted there. Normally, hunting revolved around quantity rather than quality, although Thai's revered trophies and many rural shops were adorned with antlers and horns. Prasong was an unusual exception to the rule and I had come to admire him greatly.

Often in late evening conversation he had told me of his familiarity with the great forests along the Burmese border further north where huge herds of un-hunted gaur still roamed. The fact that the jungle had no defined national boundaries made no difference to Prasong. Who cared if you were in Thailand or in Burma! When I mentioned my concern over running into a Burmese Army patrol in the frontier area, he had very casually said, "*Mai pen rai* (no problem!), we will kill them before they can arrest us!" It sounded so innocently simple! I was at that time a Lt. Commander whose soul and posterior belonged to Uncle, who I was sure would take a very dim view indeed if one of his officers undertook such an audacious and foolish trek into areas where he had no business being—gaur or no gaur!

I told Prasong about the man-killing tiger at the Isthmus that had been killed just before my arrival. He listened and said, "Wait a minute and then we will visit the ward. I want to introduce you to someone!"

Working our way among patients in beds, on the floor, and hunkered in huddles in every corner, we reached a cot with a mosquito net drawn over it. Prasong tugged the net until an edge was raised and a dark weather-beaten face with a shock of bristly, steel-gray hair looked out at us from under a multitude of bandages. When the body had joined the face at the edge of the bed, Prasong introduced me to the *phuyaiban* (headman) of a small village near Ban Huai Sai Khao, a community on the "Bangkok-Singapore highway" a few miles south. The headman could find no unbandaged hand to greet me but bravely smiled and nodded, although his pain was obvious. He looked like he had just made an overly affectionate pass at a buzz saw.

Prasong told me that *phuyaiban* Mali had been brought in the day before by villagers who felt he was about to die and felt that many problems with spirits could be avoided if he died in the dispensary rather than at home. But antibiotics and the resistance of the jungle dweller are an amazing combination and Mali was on the road to recovery, although he would never enter a beauty contest after his horrible wounds healed.

As he sat up on the edge of the cot, I could see that he was a man of unique character by virtue of the unusual tattoo on his thigh. Prominently displayed on his right leg, extending from the knee to his crotch was a huge phallus. Not just any old phallus, but a depiction of a huge penis with two legs and a tail! Mali, smiling somewhat embarrassed as I looked at the adornment, acknowledged that he felt that it paid to advertise. He felt his "macho" qualities should not go unrecognized.

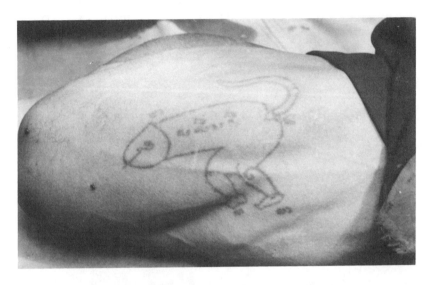

Mali was no ordinary jungle man. He displayed a huge tattooed phallus on his thigh attesting to his prowess as a lover.

Mali's troubles had started early on the morning the day before when a village herd boy had come to his house and yelled that a *demon* had killed a goat. As the *phuyaiban* , he was duty-bound to come and assist, and by virtue of his position was permitted to own a shotgun. Bound with copper wire, terribly corroded, and loaded with shells that only occasionally went off, the shotgun was the best defense the village could muster, and Mali took off, guided by the herd boy in pursuit of the *demon*.

Thailand is a paradox when it comes to jungle lore. All large cats are called *süa* with added identification of *krung* (great) for tiger; *dow* (spotted); *meg* (clouded); *fai* (fire); *and dam* (black) designating the many species of large felines that they knew such as tiger, spotted leopard, black panther, clouded leopard, golden cat and a number of smaller varieties. Surprising ignorance existed such as the belief that if the small leopard cat (*Felis bengalensis*), which is about the size of a household tabby, were fed enough fresh meat it would become a real leopard. I was often

The Thais believed that if the author fed his pet Asian leopard cat enough raw meat that it would grow up into a real leopard.

discreetly criticized when Thais would see my pet leopard cat and openly wonder why I fed it so little causing it to remain permanently dwarfed.

More amazing still was the firm belief in the northern forest areas of the existence of a huge silver tiger, who in contrast to the well-known normal tiger, was silver gray with black stripes and possessed a bristly mane and a tufted tail. Knowledgeable jungle men could differentiate the call and the spoor, but no one, at least no western biologists, could figure out what they were talking about. Ironically, I had once traded an African zebra skin to a friend who lived in the tribal areas. It soon became widespread knowledge that the existence of the strange silver tiger had indeed been confirmed after hill people saw the skin which met all the criteria of silver with black stripes, bristly mane and tufted tail. Without the hoofs or equine head as distractions from fact, it would be hard to convince villagers that the creature did not exist. After all, they had not only seen it, but had actually touched it with their own hands. So are myths made!

The *demon,* however, that Mali set off in pursuit of was a far more tangible creature. On the way to the site of the goat kill, the boy had told Mali that he was sure that two leopards had been involved. Although he had run off as quickly as he could, he felt certain that he had seen both a spotted leopard and a black one.

The latter is no rarity on the peninsula, which for years has provided most zoological collections throughout the world with their black panther displays. The Thai and the Malay governments had, more or less, a world monopoly in the international trade in black panthers since an estimated 85 percent of all the leopard in the southern jungles were black. Normally spotted leopards were a rare phenomenon. In the dark rain forests, apparently, the melanistic phase is more suited to survival, but black and spotted cats do interbreed with half the offspring being black in color. Black panthers bred to a mate of the same color always produce all-black cubs. If the observations of the herd boy were correct, it appeared that a black and spotted pair of cats were temporarily hunting together since it was the mating season.

Mali was a bit apprehensive, he told us, as he reached the open field where the goats had been attacked, because in nearby villages within the past week two other frightening incidents had occurred both under similar circumstances. A herder in an adjoining hamlet had also seen two leopards who had been very aggressive in their attacks on his goats late one evening when he was returning to his village. In a bushy area only a few hundred yards from the huts a large leopard had leaped on the lead goat and with one quick bite had killed it and proceeded to drag it away. The herder, a grown man, had thrown a stick, and the leopard, a huge shiny black panther with green eyes, had snarled and spat at him. For a moment he thought he was going to be attacked also, but in his anger at losing a goat he had thrown a second stick and the panther had turned in a fluid movement and re-entered the bush leaving the bleeding goat on the trail. It was then that he saw that the huge cat was accompanied by a second leopard that had remained hidden. With two leopards near him, his courage had rapidly dissipated and he had run for the village with his terrified goats close on his heels. Soon the entire area, for several miles around, had heard the story because news on the jungle telegraph travels quickly.

On the following day, the same scenario was replayed at a rice paddy about two kilometers away. A ten-year-old boy had gone out with his goats accompanied by a couple of village pariah dogs who normally spent most of their time lounging around the temple grounds. It was a misty morning with the air thick and humid. Suddenly the dogs had bristled and yapped in unison at something unseen in the bush. Not known for their aggressive characteristics, the dogs had been undecided on what to do, and the herder had, in his curiosity, joined them. Bolstered by human backing, one of the dogs had entered a bushy mound of dirt on the paddy embankment. Suspecting nothing dangerous and assuming that the dogs had cornered a palm civet, the herd boy tossed a stone which brought a low growl from the bush. Before dog or herder could absorb the significance of this, the leopard had sailed through the air in a flashing attack on the nearest dog which, unfortunately, made a nimble jump to the side in its attempt to get behind the boy. The rapidity of the attack left no time for the herd boy to do anything but stand rooted to the spot paralyzed with fear. The leopard had been the spotted member of the duo and had raked a slash across the boy's thigh, ripping his pants from his body. He had been bowled over and felt sure that momentarily he would die. The leopard had jumped past him after its lightening-like attack to futilely chase after the yapping, hastily-departing curs. The boy had lain still and watched as the leopard's mate joined it in bounding off across the paddy field toward an adjoining jungle patch.

Bleeding badly, the frightened herder had left his goats to fend for themselves and had hobbled home as best he could. Because of the extreme danger of septic infection from even a minor leopard scratch, the boy had been brought to the dispensary for treatment. Gossip in the coffee shops again centered on the unusual aggressive behavior of the pair of leopards. Jokes were passed around about the effect that sex, or lack thereof, brought on such aberrations, because the cats were obviously a courting pair. More laughter ensued about wishing that the leopards would concentrate on the purpose of courting and leave innocent people alone.

Few villagers were concerned about man-eaters because for some unusual reason leopard in southeast Asia rarely take on the habit of eating people. In India this has been a gruesome practice of long standing and some man-eating leopard there have attained historical notoriety. In Thailand, as well as much of the rest of southeast Asia, tiger occasionally turned man-killer or man-eater, but not leopard. Even the taking of village goats was an unusual occurrence and, in spite of a fairly dense leopard population, the village people rarely saw the animals and were seldom bothered by them.

Mali and the herd boy soon found the place where the goat had been killed and, since there was little cover nearby, they focused their attention on a similar little mound on the paddy edge from which the other leopard attack had occurred. They found the remains of the goat which had by now been fully consumed by the pair of leopards except for the head and hoofs. A few pieces of skin were also left behind.

Deciding that the leopards had left, Mali and the boy went to a stream about a hundred yards away to get a drink before walking back to the village and attempting to round up the scattered goats on the way.

They had just entered the bush fringing the small water course when, without warning, a leopard had leaped from the bushes at a distance of less than ten feet and knocked Mali flat on his back. With the panther directly over him, Mali had held his arms across his face in hopes of diverting the jaws of the cat. The leopard had his forearm in his mouth and Mali could feel the fangs crunch against his bones with every crushing bite. The leopard, in its initial spring, had given Mali's head a raking blow above the ear which had ripped his scalp loose in a large, profusely bleeding flap which partially blinded him. The leopard had flailed away at his prostrate victim with all four feet causing huge lacerations. Then, almost as quickly as it had begun, it ended and the black panther had disappeared. The herd boy had been a mute witness to the entire event feeling any moment the spotted leopard was going to kill him as well. Although the other leopard had most likely been right there, it had not participated in the unprovoked attack. Mali, with the help of the herd boy, managed to get back to the village from where a small rickety bus had taken him to Krabi.

He grinned as he told me his story, and affirmed how happy he was that he was alive. In a part of the world where cosmetic surgery did not exist, Mali would, for the remainder of his life, bear grotesque scars and a horribly mangled forearm that probably would never again be fully functional. But, oh, the tales he would be able to tell when the family sat around the fire on long evenings back in the village! The attack would be relived a thousand or more times becoming more vivid with each telling.

Prasong asked if I had time to join him to try to shoot the leopards since he was planning to visit the village. The courting phase for the cats would soon be over and the pair would split up. It seemed like a good possibility and well worth the try. I had two extra days and was particularly enthralled by the fact that one of the cats was spotted. Now this requires some explanation since I have been trying to explain it to myself ever since!

The incident at Krabi was during a period of my life when I was hunting strictly for pleasure or, as in this case, for necessity. I had still to hear of record books, and about rare and coveted trophies. That a black panther fell into this latter category still escaped me entirely. In my mind a leopard had spots, and since in the area I lived few leopard were well marked, I, in my naive thinking, felt this was the only leopard worth shooting. Black panther, I felt, was so ubiquitous that it hardly seemed worthwhile. Knowing now what I obviously did not know then was that a rare opportunity existed for me. In my misdirected priorities I looked forward with anticipation to collecting an honest-to-God spotted leopard, paying hardly a second thought to its jet black mate. Prasong and I quickly gathered our guns and gear and headed south on the road to Huai Sai Khao.

On the way, Prasong pulled off the road in front of a disheveled, thatched roofed hut set among some banana and papaya trees and hailed an old man sitting on a stump in front of the hut. The man waved but looked right past us as if he were ignoring our presence. On closer approach, I saw that the frail, little old man had totally opaque, sky blue eyes that stared vacantly, yet mysteriously, into space. It took a moment to realize that he was blind from cataracts. He was dressed in a soiled, gray, toga-like garment that was most unusual for the area. Prasong introduced us and then told me that the man was a *maw phi,* or spirit doctor, who could divine the future and control destiny. Few major undertakings are ever initiated in Thailand without first seeking such counsel, and I should have realized that Prasong would not have considered tackling a leopard with me before first finding out if the moment was propitious—and advisable! I can now recall few of the observations made by the shaman as we discussed our plans with him, but I do recall that he gave his blessings to the idea and, holding my hand, said that I had a very long life before me. I was immediately convinced that his powers were flawless! He even went on to say that I would live until I was ninety-two. With these prospects of a few more years of life now firmly secure, Prasong and I gave him a few cigarettes and some

The spirit doctor examined my hand and prophesied success in my hunt and added that I would live to be ninety-two.

coins and loaded back into our vehicle, leaving the old man staring, wide-eyed, and perhaps seeing things in the spirit world that we could not see.

Mali had given us a note to one of his brothers and having contacted him we bought three goats which we intended to tie out as bait. The brother informed us that everyone had heard the leopard growling around in the bush the previous night and we laughed thinking that maybe the courting process had gotten down to basics and that the leopard were hopefully now less mischievous. We did hope though that they had worked up a good appetite.

Two villagers accompanied us to the area where Mali had been attacked and we searched for a good stake-out site. The stream bank was much too dense and we quickly eliminated that idea. The paddy fields were quite open, interspersed only with occasional huge sugar palms. We looked at these and also dispensed with the idea of perching on one of these with no branches and an elevation far too high to even consider a shot. The bush areas were sparse and only about head high around the cultivated plots.

On one of the dikes surrounding a paddy field was a fairly elevated knob with a cluster of dense brush on top. Such mounds were visible in several places and were the result of dirt pile-ups resulting from leveling the paddy and then constructing ridge dikes around the field to hold water in the rainy season when the rice was growing.

We decided that this offered us the only clear visibility for a shot and at the same time gave us some security by being in an elevated position of some four to six feet with bush to hide in. We tied one of our goats in the paddy about twenty-five yards in front of us.

Prasong had a unique and novel method of making the bait goat bleat to attract the leopards. He took a paper stapler from his pocket and clipped a staple through the ear tip. Since ears are made of cartilage, I am sure the trauma was minimal when one considers the ear mutilations by stockmen in America from notching and ear tagging of cattle. The sting must have been adequate, however, because our goat, finding itself alone, started a continuous calling. Ear staple or not, I felt it would have kept up a constant calling, hoping that someone would soon notice that it was not in its home corral, where it knew it should be, and would soon come to retrieve it.

Prasong and I made our way to the hummock to settle in for the night. Just as we scaled the edge a rustling scampering occurred directly in front of our noses. I know we both thought "Oh my God, we picked the one place to sit where the leopard was hiding and waiting for us!" If there was ever a cause for instantaneous panic, this was it! We were rooted, it seemed, in mid-air and were pointing our guns at an unseen danger and staring at vacant space. All was now silent. Prasong turned to me with a quizzical look and shrugged. I shrugged back. Then we heard a rustle again and saw a huge monitor lizard slide away down the far side of the hummock and waddle in all haste across the open field. We both felt greatly relieved that we had not had to face our leopard at a distance of three or four feet!

The possibilities of taking a leopard under the stakeout we had arranged was rather remote, but in the enthusiasm of ignorance we sat side by side certain that soon our quarry would show up.

It was a warm night and mosquitoes serenaded us with a loud whine at a furious and continuous pace. They had never had it so good and seemed determined to eat us up before the leopards even had a crack at us.

A pale moon came up about nine o'clock and we began to be able to see our surroundings a bit better. I had a two-cell flashlight that I had practiced with by holding it alongside the rifle barrel and I felt that in the distance of twenty-five yards it would suffice quite well. A bit earlier I had looked through my scope set at 2X and pondered about where all the fabulous advertised light gathering qualities of the optics were, because I could not see a bloody thing!

I had once sat up for wild boar on a similar night. When my fellow hunter had nudged me to tell me that the pigs had arrived and that I should shoot, I recall putting up my rifle and seeing nothing but black blobs in the sight picture. I was convinced I would make a killing shot on a bush or a boulder because nothing in the scope looked like a boar, sow or even a piglet! I shot anyway and amazed everyone, myself most of all, when I found I had brought down a nice tusker with a perfect neck shot. It is hard not to say, "Yup, that's the way I do it every time!" I hoped this night's experience with the leopard would go equally well.

While such thoughts went through my mind we both suddenly realized our goat had become silent. Prasong and I nudged each other and we both stared intently at the goat to see if we could see what it was looking at. I think we both got the same jolt when we saw that the goat was staring directly at us! It took a second or two for the realization to sink in that the goat knew we were not its problem and that whatever it had become aware of was coming from our direction. We could hear nothing. Neither of us relished the thought that maybe the leopards were not only approaching, but might attempt to utilize our hummock as a vantage point to launch their attack. How the leopard would react to finding us already entrenched on the spot was not a pleasant thought, especially since the cats were approaching from our rear and we did not dare attempt to maneuver an about face.

Several moments went by while each of us was convinced that the thumping of our heart could be heard by every animal for miles around. Suddenly there was a rush of sound next to the hummock and a scuffle by our goat and then all was silent again. We could see movement but nothing else distinctive enough, but we were sure it could have been nothing other than the leopards and that our bait had been taken. We hoped the sacrifice was not in vain.

We had agreed that after the goat was killed we would allow a few appropriate moments for the cats to settle down and that I would then snap on the flashlight. The plan was that we would both then fire at once, hopefully not at the same target. So much for well-made plans.

I slowly raised the rifle to my shoulder and put the flashlight in my left hand so it would point at least in the same direction as the muzzle. I pushed the button quietly forward expecting any second now to have a pair of startled leopards with staring eyes looking into the beam of the light. I pushed harder and got a bit of a shock when nothing happened. I tried quietly to shake the confounded gadget and again pushed the "on" button. Again nothing! By this time Prasong was having apoplexy not knowing what I was doing. I rattled the flashlight again before giving it up as hopeless, and then we both realized that if we were going to shoot we'd better do it fast, even if we could not see what we were shooting at.

I have often admired the hunters who write about killing animals by starlight. By starlight I can barely see my hand in front of my nose, much less try to successfully kill a dangerous animal. I tried to remember my successful shot at the boar and thought that maybe anything was possible with a bit of luck, and besides, the *maw du* had said I would live to be ninety-two and that was still a long way off.

We raised our weapons and attempted to sight in on a recognizable target. With the goat prostrate, it at least narrowed the field somewhat—if it was standing, it must be a leopard. We both fired at virtually the same time. There was a loud roar, thumping sounds of something running and then silence. We felt confident that we had hit our visitor; but since something had run away we could only assume that only one had been hit because we could see a lighter body already lying near the goat.

Not wanting to get down in the dark without a light, Prasong attempted to create a torch while I fidgeted with the electronic marvel called a flashlight. After tightening the bulb, stretching the internal spring and rubbing the battery ends against my shorts, I tried it again and almost dropped it when the light went on! We quickly directed the light toward the goat and there, within five yards, lay a beautiful leopard. I was ecstatic! It was covered with proper spots the way everyone knew a leopard should look. It was the male of the pair.

We decided that since the black female had probably been almost invisible to us in the darkness that we had both obviously zeroed in on the lighter spotted cat. We cast about for blood that

The large spotted male was shot by Brandt while its jet black mate escaped the ambush.

night and again the following morning, but found no other blood or the spoor of the black panther.

I had to leave the next afternoon but couldn't have been happier with my well-salted skin in the back seat. It would adorn my wall. What would I have done with a plain old black one anyway? So much for stupid rationalization! But my last meeting with the black panther was still to come.

About ten days later, I was in a provincial capitol called Nakornsithamarat where a Chinese animal dealer was headquartered. I often stopped by because he had an endless variety of newly-caught jungle fauna constantly in stock from spotted baby tapirs to brilliantly-marked Argus pheasants. As I walked about among his sturdy but horribly small cages, I jumped as something hit the barred front next to my leg with a blood-curdling growl. I could see by the huge claws wrapped around the bars, trying to reach me, that it was a black panther.

The owner came to me and said the caged female panther had been trapped by some village people at a place east of Krabi. He didn't remember the village exactly but I felt quite confident that it was the same cat Mali, Prasong and I had met before. It had lost one canine tooth in attempting to extricate itself from the trap. The villagers had received the equivalent of $150 in U.S. currency for the animal.

I asked the entrepreneur what its fate would now be. He said that its value was considerably reduced because of the broken tooth and inquired if I knew anyone that might want it. He would flip it a cyanide capsule and have the skin immediately available. I declined the offer, still pleased that I had taken the spotted leopard. Black panthers just didn't look right.

As time went by, I often wished I could have done it all again. One can only kick oneself just so many times for having made a foolish decision, but I no longer ask myself or anyone else, "What would you do with a plain old leopard with no spots anyway?"

RILE UP A LEOPARD

by

Gordon Young

Karen tribesmen request help in removing a predatory leopard that almost succeeds in turning a routine stalk into a near disaster for the hunter in a thick, thorn-studded jungle.....

We were hungry, looking for a campsite which should have appeared around the next bend, or the next, and we hadn't taken a break in two hours of exhausting rock-to-rock, going up the twisty, mostly dried-out Mae Seudh stream bed. Then, just at high noon, the distinct, deep bass "coughs" echoed down the canyon from no more than what seemed a hundred yards upstream. It was an unexpected thrill at that time of day. It did marvels for my growling stomach; hunger became unimportant, forgotten.

What was inducing a leopard to sound off at noon, in broad daylight? In my book, leopards were supposed to be napping soundly at such a time.

I exchanged confused looks with my Lahu companions, all tribesmen from the mountains of northern Thailand. We hunkered down along the right side of the stream and I slipped off my light rucksack. My hunting companions were as baffled as I was. They looked up the stream, heads cocked, questioning and obviously amused as well. It made about as much sense as a rooster crowing in the dead of night.

I had come to the Mae Seudh area of far northern Thailand at the request of Karen villagers to kill a large, aggressive leopard that had several times brazenly entered the villages to carry off dogs and had regularly killed livestock. Sitting in isolated field huts, guarding their crops, the Karens had several times felt threatened as the leopard growled and coughed in the night as they huddled around their small fires. Unlike the Lahu tribesmen of the north, the Karens are far more timid about hunting an animal as potentially dangerous as a marauding leopard.

Seated beside me was the young chief of my Lahu hunting companions, Javalo, who smiled and whispered to me, "You come to this valley where there is said to be many leopards, so they mate even in the day. It is the sound of calling out to a female hiding someplace . . . he gets the scent but can't find his teasing mate . . . he has to relieve strong desires . . . ."

"A mating call, yes, that's what it is," I whispered back, having listened to the various and wondrous vocalizations of leopards with fascination since I'd been a boy. It didn't surprise me that Javalo could know so quickly that it was the voice of a male and not a female announcing her receptive heat. Javalo knew everything of the jungles.

Maybe I had been missing out on leopards, but it had been the first voicings I'd heard in the daytime. In the higher mountains, where we did most of our hunting, leopards called from dusk into the dawn, but always became silent after sunrise. In the mountains of northern India I'd often listened to leopards calling; it had been the same pattern, invariably dusk to dawn.

We waited a few minutes checking the breeze coming downstream. The leopard sounded again, getting closer. He wasn't getting our scent or hadn't heard the whispered sounds we had made. I liked the sound of the voice. It was deep and strong, surely a big male, as Javalo had guessed immediately. I motioned to my five companions to remain where they were and stepped off up the stream, leaping as quietly as I could from boulder to boulder, expecting to see the leopard even before I got around the slight bend. The wind continued to come down the valley, no question about that, and it cooled my sweaty face. I just needed to be silent and quickly get to a good ambush point. Get around the first bend then sit very still off to one side and let him come into view. I made it before the leopard spotted me and squatted on a large boulder with my rifle ready.

I carried my old converted Army Springfield 30/06 with 180-grain silvertip. The minutes passed by very slowly. There were no more rasping calls from the leopard since I'd left my gang of Lahus. The leopard had to be very close now, if he hadn't gotten my scent and made a sudden retreat away up the valley's side or back upstream. Nothing indicated what he might have done at that point. I decided after another five minutes that if he was pausing suspiciously there was a possibility that I could force his hand with a little decoy trickery. Curiosity had killed other cats before. I put the palm of my hand to my lips, sucked lightly against it to produce a sound like a distressed rodent, a favorite trick hunters use, perhaps all over the world, designed to turn the head of any predator. It wasn't the first time I'd attempted to decoy leopards. They could care less for the bite-sized rat, but the curious sound was usually irresistible even to a tiger.

One series of calls. Nothing. Another series of calls. Nothing again. I was about to give up, thinking that I had blown whatever chances I might have had. I sat on the rock weighing out the wild chances that might still be there.

For some reason, I thought just then of a leopard, one of the first real declared killers that I encountered, even if I wasn't the hunter to kill it. I was a slender lad of thirteen, hunting with my father on the open badlands of Tang Yang in north-central Burma, several miles to the north of that town. A rather notorious leopard was reported by Shan villagers to hole up in the dense brush of the canyons that drained the red lateritic slopes of the plateau. It was a leopard that had killed a number of Shan cattle and many pigs; and was suspected as the probable killer of a woman and a small girl in the recent weeks before we stopped at Tang Yang. We thought it would be simple: we could probably flush the leopard by throwing rocks into the canyon, with my dad and I waiting from the rims with our shotguns loaded with L.G. shells, the British equivalent of our less-used, six-pellet "0-Buck."

Conducting the rock-throwing were several of our wranglers, Wah tribesmen who traveled with us in those days before motorable roads went beyond Tang Yang and the China border. We seldom had Wahs with us on hunts because Lahu tribesmen, who were the renowned hunters in the land, usually were there to accompany us. The Wahs, famed warriors, more typically disdained hunting, unless it was for human heads. But the Wahs happened to be there that day, so they volunteered to pick up the task of beating the bush, probably because they wanted the diversion.

After numerous tries at it, we decided to work the different branches of the canyons by waiting at the heads and having the rock-throwers come up the canyons in short sections, with the intention that if the leopard was flushed he'd run up the canyon to where my dad and I waited. The throwable-sized rocks apparently got scarce, so one of the Wahs named Sam-pluk decided (demonstrating typical Wah warrior mentality) to disregard his dad's cautious admonitions to stay up on the canyon rims. Entirely on his own judgment, he went down into the canyon where, obviously, he could beat the bushes more effectively. Why should a Wah warrior, even a reformed headhunter like himself, be worried about a cat that weighed, probably, no more than his own 140 pounds of chunky muscle? What kind of a Wah would he, Sam-pluk, be if he showed fear? Fear was something Wahs didn't recognize: men joked about it because it was for the non-Wahs of this world; fear was the silly stuff of aged women and small children.

I wasn't aware that the man had gone down into the canyon until the moment when the leopard gave two choppy coughs in the thickets just below where I happened to be standing. This was a portion of the canyon out of my dad's view from where he stood on the opposite rim. The leopard broke out from his hiding place and pounded off in the wrong direction, down-canyon toward where I was suddenly very amazed to see Sam-pluk step into an open spot at the canyon's bend! It happened very fast. I didn't have a prayer to get a running shot at the briefly-visible leopard flashing through dense ferns and brambles. The approaching man was directly in the leopard's path. In one flowing leap that blurred his spots with the yellow of his fur, he went straight for Sam-pluk's throat. I might have shot just then, but, fortunately, I saw Sam-pluk out of the corner of my eye just before I was about to squeeze off the shot that might have hit the man as well as the leopard!

Perhaps I thought that day of that incident of a dozen years earlier in my life because it always comes to mind when I think of or encounter leopards, even in the zoo. I witnessed the unforgettable ferocity and lightning speed of leopards, which forever I would respect. And at the same time, the amazing skills that are part of some men's warrior training, which, when a man must, has to come by *trained* reaction—he hasn't time to be thinking about what he's going to do. Famed as a swordsman—as much so as the Kachins of Burma—a Wah is instantly reactive with his blade, suggesting something of the arts used by Samurai warriors of ancient Japan. In any event, it was my privilege to see two movements of Wah martial arts that are stored there forever in my memory: the arm, hooked protectively out in front of his throat, offering an elbow instead of a vital exposure; and the highly practiced swing of his *dah* (long-knife) which he drew and flicked at the same time. I saw all this in a glimpse that I had to recreate later to understand what happened; this had to be sorted out of a very fast mixing of man and leopard. The leopard hit the man with all four feet forward, then seemed to bounce back off like he'd hit a springboard, and as though spraying from a hose, his blood made streaks in the sand. I stared incredulously at the spurting blood.

Small wonder. The leopard's head remained attached to its body only by the shreds of some skin and muscle on the right side of its neck! Sam-pluk's blade had severed at least 80 percent of the leopard's thick neck with a single, slicing swing of his knife!

But now, back on the Mae Seudh, my legs were getting cramped. What was there to lose? One more series of squeals . . .

The leopard suddenly flashed into view as he stepped all too eagerly around a large boulder. He saw me immediately when I moved to bring up my rifle; a very subtle move on my part, but he saw it at once. He turned too fast for me, leaping to his left to move out of sight behind a dense stand of nettles. I led out slightly ahead of him and loosed the Silvertip straight into the nettles in a line calculated to meet his leap.

It seemed that I had him and it didn't surprise me. There was a furious gurgling roar, a heavy thrashing around in the nettles, then to my complete amazement, he bounced right back out of the bushes onto the sandy spit in front of me. He was on his haunches and pawing at what seemed to me to be his neck. I was wrong about that because he was actually working at the sudden injuries he had sustained on his muzzle. I had my sights on him then, deciding to place my second bullet into the slight ruff of his neck. Certain that I'd hit him fatally, I lowered my rifle to watch him collapse without tearing up his magnificent hide any further. I was positive just then that I'd gotten a beautiful male leopard with one clean shot: the cleaner the kill, the easier the tanning and processing.

I was way off in my assumption, still learning that you never, ever take anything for granted when hunting. Perhaps more especially so when dealing with leopards. What had actually happened was that my bullet had hit a hidden boulder just in front of his head as he went through the bushes and, splattering off the rock, pieces of the disintegrating bullet sprayed the leopard's muzzle. One piece of the 180-grain Silvertip knocked out one of his canine teeth; other small chips sank into his lips and nose. His sudden, angry reactions were understandable. It was one of those miserable errors I hoped would never come along to interfere with my hunting. There are many other ways to rile up a leopard and quickly make him or her far and away the most dangerous animal one will face in the jungles. After all, they have all the potentials: speed, cunning, fury, and, very importantly, lethal size.

But this, it seemed, had been an open-and-shut case. No, I told myself as I lowered my rifle, another shot would be wasteful and unnecessary. I could simply watch what would be one of the easiest leopards I had ever killed sink down on the sandspit and die, in a clean kill that my Lahu companions would admire. It was ridiculous, but I kept thinking about keeping the damage to only one hole in the skin so there'd be no need for extensive repairs. I felt smug, inwardly congratulating myself on my luck.

The leopard suddenly snapped out of his preoccupation with his wounds and, after rolling on the sand a few times, leaped for the same bushes he had tried to escape into before. In my haste, I reacted with even more stupidity this time, guessing rather than aiming when I fired a shot into the bushes in the vicinity of what I hoped was his moving torso. Somehow I hit him again. There was a loud, coughing growl and I thought he'd bounce back out again as he did before. My shots seemed to be confusing him as much as he was confusing me! But I knew by the thrashing sounds in the bushes along the right bank of the stream that he was moving quickly away up the stream this time.

Then it was all silent. My smugness went flying; I felt very bad and embarrassed. I didn't feel like explaining my mistakes to my Lahu companions who liked to exaggerate big praises for good shooting and had discomforting, silent moods to display for hunters who lost control. I had heard them say often enough among themselves concerning one of their own people, "The man should have his eyelashes plucked! Easy shot and he missed!" I tried to console myself. They were *not* easy shots. No, but I shouldn't have shot at all!

Upstream, some fifty yards away, I caught a faint growl or two and heard a rock click against another rock. The leopard was moving fast, he could not have been very seriously wounded. I wondered why he had made such a big fuss out of my first bad shot, not knowing at that instant what my bullets had actually done. I turned around to find Javalo right behind me, holding my double 12-gauge shotgun. "Where does he lie?" he grinned.

I wiped the sweat off my face and swatted at the numerous little hovering bees swarming around me. I didn't want to say anything. Then I said sheepishly, "Stupidity took over instead of brains. Ought to have my eyelashes plucked!"

"You hit him, there was a lot of complaining, then a second shot."

"Aye," I said, and explained bitterly what had happened.

Javalo nodded grimly as he handed me my shotgun and two extra rounds of buckshot. "It's loaded. You won't need more than one shot," he said, much too confidently, then added, quoting from a Lahu adage, "Who riles a leopard, follows a leopard." It was as important a code with Lahu hunters as with careful hunters around the world that a wounded animal, especially a potential killer, never be left to attack another person. "You've got a good cover of heavy lead with the shotgun—one bullet in the rifle is easy to miss with. Be alert, he'll come out fast!" I didn't like the cheerful sarcasm in his voice.

"I promise you, I'll be alert," I said, irritably, trying not to appear nervous. I exchanged guns with him. "Don't explain to me about the differences between rifles and shotguns!"

Javalo shrugged and looked amused. "If you have trouble tracking, whistle and I will come to help you. But you should be alone and ready to shoot fast, *in any direction*! And, what's more, I don't want to be in the way!" With that, I thought he'd burst out laughing.

I nodded, glaring fiercely at him, checked the pair of double-ought buck cartridges in my shotgun and went over to the bushes where I'd last heard the leopard moving. I hunkered down to inspect the first splashes of blood and listen carefully for a few moments. I also needed to muster more courage to proceed. I was breathing much too hard, even without real physical exertion—I had to get a good hold on myself or . . .

Safety off, ready, I stepped out at a crouch. I determined that I must move slowly, keep my eyes mainly up, glance briefly at the blood trail that was as plain as anything I'd followed. My brain seemed full of leopard tactics just then from the hundreds of campfire stories I'd heard. Foremost was the fact that wounded leopards will end up in ambush. Watch carefully! "You won't see him first. He's not a tiger who'll growl loudly before he leaps for you," I recalled one hunter saying. A man has one bat of an eye to spot and shoot. I didn't like the narrow odds. I'd been very lucky on two other occasions before when I went in to "walk out a wounded leopard" only to find them stone dead. One had been in the dark and I used a flashlight. I thought of how vulnerable I'd been then and what stupid things a "responsible" man with a growing family will do on hunting trips to jeopardize himself and his family.

The leopard was bleeding quite profusely, mainly from the right side where bright red arterial blood splattered off to one side. It seemed that he'd bleed to death if I waited. Then I decided that I had to keep him moving to insure that he didn't rest long enough to stop his bleeding. It was what Javalo would have advised had he been alongside just then. I tracked up the stream's side for some fifty yards, pausing every few steps until the trail turned into the stream where there was a small pool of water. It was red with the leopard's blood. He'd laid his wounded side in it, so I could be sure now that my second shot had hit him somewhere on his flank or limbs. He then got up from the pool and crossed the stream to go into a dense thicket of thorny mimosa about five yards on the other side. An alarm went off immediately in my brain: *Watch out! Here is where he intends to ambush!* He would want me to get down into a crawl to go into the small tunnel he'd created from the edge of the thorns and while I fought to get through, he would make his fatal pounce!

As though the spot were a mine field, I backed up to the stream and very carefully skirted around upstream until I was past the spread of the mimosa thicket. Then I squatted down to peer under the thicket and scan the areas behind it. I had a very strong feeling the leopard was waiting for me just inside the thorns, *knowing* that it would be difficult for a man to crawl through it. Seeing nothing in the relatively clear floor under and behind the mimosa thicket, I moved around to look for the blood trail at the back of the thicket along a flat area about ten yards wide just before the abrupt rise of the canyon's wall. I spent five very cautious minutes there, trying to pick up the blood trail on the carpet of fallen leaves.

Nothing. Why the hell not? He was bleeding like a stuck pig just on the other side of the thorny bushes. He hadn't climbed a tree because there were no more than shrubs near the mimosa thicket. So where does a bleeding leopard disappear to so fast? I was sweating profusely, my heart pounding. Now I had lost him and couldn't be sure in what direction the leopard might be lying up for me. Logically, I told myself, he'd pull off into the densest part of the thicket, stay on the flat stream floor, so badly wounded he wouldn't want to attempt to climb up the steep canyon wall. But that was, of course, a logical direction for him to take. I looked up into the patchy bamboo growth and numerous boulders above me, studying every suspicious shadow that might be part of the leopard. Nothing, not a sound other than the high-pitched buzz-sawing of cicadas. I sensed the leopard's presence not far from me and *knew* he was lying flat somewhere within the 180 degrees from where I crouched, soundless, and not about to sound-off before he made his leap at me.

I soon convinced myself that the leopard wasn't on the flat area around the mimosa thicket. He had to be up on the side of the canyon. So I faced the canyon, my back to the direction he had entered the mimosa bushes. The silence continued, more ominously than ever. After a few moments of careful scanning of the numerous possible hiding places above me, I heard Javalo whistle gently like a quail. I whistled back. He whistle-talked (something that is easy to do with the Lahu language's tonal and monosyllabic nature). I whistle-talked some more . . . .

The leopard was well into his charge before I saw him. He began without a sound, then coughed out a terrible blast of what I later decided was from pure agony and pain mingled with his anger. I could not know just then that he traveled on three limbs, a factor that probably allows me to relate this story. He moved with such speed, in spite of this, that I was clutched by terrified wonder. The sound is hard to describe, a combination of inhaled, exhaled, hoarse, spitted growls. He was a flashing blend of brown, his spots and ground colors blending. I had my gun up, but he was moving very fast with twisting bounds around the boulders, zig-zagging down the slope, an impossible target. Then from about thirty feet from me, he leaped right out over a large mass of rock and I was looking up at his pale undersides, his limbs spread.

The Mae Seudh leopard who set up an almost perfect ambush to kill his pursuer at thirty feet.

My right barrel roared and I had enough remaining presence of mind to step off to my left as fast as I could. The leopard crashed to the ground where I had been standing. A hunter's nerve is an odd thing, one second the cool holds and he's doing great things to save his ornery hide; in the next instant, that nerve escapes in most undignified ways and he's running like hell to save his ass! I was in just such a hurry after that first shot because I could picture the leopard, which I might well have missed, twisting away from his fall and cat-acrobating with one more leap to slap me down. I turned in my rush to see that he was writhing on the ground where he'd landed, so I stopped, turned and fired my left barrel into his neck.

This, the angriest of leopards I've ever encountered, probably would have had me had he been able to use all four of his legs. As it turned out, my second shot with the rifle had caught him on his right forearm, shattering his elbow joint. Even on three legs, he'd been a streak! I will never cease to marvel. I doubt very, very much that I could have hit him with the shotgun had he been moving even slightly faster. And I have never been able to figure out why I lost his blood trail after he went through the mimosa thicket.

However, there is an important epilogue to this encounter on the Mae Seudh that is far more important than that the very large male leopard, despite the bad tears in his hide, became a beautiful trophy skin. This portion remains with me forever.

Javalo, who was behind me just on the other side of the stream, heard sounds that motivated him into actions that touched me deeply. From his perspective, there had been a quick groan or two from the leopard, a shot, a crash and increased growlings of a more blood-curdling nature, then one more, perhaps a last, desperate shot. His interpretation: I had had it! And so, with a great war cry, the man who would have killed the leopard, had I not done so, came crashing through the mimosa thorns. He held his long-knife in his hand instead of my rifle, ready to mix it up with a leopard that was tearing me to pieces! He was simply magnificent and was probably timely enough to have saved me had that been necessary. When he saw that I stood there near the leopard which had now quit its last few twitchings, he laughed loudly with joy and pointed at the leopard, nodding approvingly. Then his face twisted into frowns of pain and, sheathing his long-knife, he ran

The beautiful, dark pelt of a large, deep jungle leopard that almost cost Young his life.

his hands over his bloody head and arms. Through his grimaces he said, *"Ai-ya-ya-ya!"* I realized only then that in his haste to reach me and assist me, if needed, he'd charged through the wickedly awful mimosa bushes and the thorns had raked his head, ears, shoulders and arms mercilessly. He bled from many abrasions where the rasp-like branchlets had caught him. I could only imagine his pain, and I knew that they would get worse a few minutes after such thorns bit into a man's hide.

"I heard the growls," he fairly shouted, breathing hard, "My spirit fled—thought you were being mauled! But happy it is to see you scratched less than me!"

I went over to him and looked at his thorn wounds. They were bad, but it was perhaps merciful that mimosa thorns are not like those of the various *rattans*. It was like Javalo to point that out when he observed, "Ah, but where would I be had that been a *rattan* thicket!" Then he burst out laughing, shouting out to the boys behind us not to come through the thorns as he had done.

"Where would *I* be had it been *rattans* and had I been struggling my last under a leopard," I laughed back. I knew that Javalo would have tried just the same, even if the *rattan* thorns killed him. It portrayed a loyalty that I shall never forget.

The leopard had been in mid-air when I shot. He had been so close that wads from my buckshot were later found embedded in his chest where my first shotgun round had hit him. My second shot, fired from about five yards, was not close enough to introduce the wads into his neck. I humbly admit that there was plenty of great, good luck behind that quick, barely-aimed shot as the leopard flew through the air at me! Later, as I prepared to skin him, he measured seven feet four inches "between pegs" and was one of the largest males I have ever taken in southeast Asia. The Karens of Mae Seudh never again complained of a marauding leopard.

There have been numerous other leopards in my hunting experiences. No two were alike, especially where the hunting was complicated enough that I could expect trouble. On the Mae Seudh, however, will remain the ghost of one very unforgettable leopard that I had the poor grace to rile up in broad daylight. My reward had almost been that ticket to the Happy Hunting Grounds.

Ah, but to find a time and place in which to make such mistakes again!

SUKANTI'S KILLER

by

Pat James Byrne

as told to

Capt. John H. Brandt

BRANDT

A man-eating leopard that has killed six villagers chooses for its seventh victim a vivacious young girl named Sukanti who disappears without even a cry for help.....

It was a warm, humid evening as the small, dark-haired girl, named Sukanti, stepped outside her hut. She was twelve years old and already quickly approaching womanhood. She quietly told her mother that she would be back in a moment and walked off into the sparse bush near her home in the small village of Dangasurada.

Her family continued making supper and, as everyone crowded around the mat to eat, comments were made that if

Sukanti didn't return quickly that she would miss her meal entirely. Several jokes were made about her perhaps meeting a boyfriend some place. Everyone knew that this was not the case and, as the minutes passed and the deep tropical night settled in, the family became more and more concerned over her absence.

Sukanti had only stepped outside her hut for a moment before she was carried off by the man-eater.

Every few moments someone would go to the door and shout her name, but no response came. The only sounds to be heard were the cows feeding in the yard and the night wind blowing in the trees. Their shouts were noticed by neighbors who came over to inquire about Sukanti's mysterious disappearance. Soon everyone was thoroughly frightened and no one moved from house to house unless it was in a group to bolster everyone's confidence.

Fear on the part of the villagers was entirely justified since in the past several months a total of six known persons had been carried off and killed. Two victims had been taken in the village of Kirama and one at Bandiri. More recently, three additional people had been killed in the vicinity of Jhaluguda. The possibilities of additional unreported deaths among travelers or itinerant wanderers might also have occurred, but no one would ever know about these additional unfortunates. Until now, no one knew if the assailant was a tiger or a leopard, but everyone knew that some death-dealing creature had entered their environs. Fear was rampant.

As the hours progressed and the wailing of the women continued, several men agreed that they should form a group to search for Sukanti. Some even suggested that this should be initiated immediately in the event that she might still be alive— although no one really believed in that possibility.

The men gathered several flashlights and some lanterns and armed themselves as best they could with clubs and axes. No one among them possessed a firearm, but everyone felt that if they stayed close together in the night they would be reasonably safe.

The leader of the group, a man named Mandolo, had only gone a few yards in the direction of the rear exit of the village when several drops of blood were observed on the hard-packed ground. A noticeable hush fell over the group. Someone suggested that since little could be done in the dark it might be more advisable to wait until morning light to continue the search. Everyone quickly agreed and the group returned to the village. The doors that night were firmly shut and bolted. The continuing murmur from the huts throughout the remainder of the night indicated that few people were asleep, and virtually everyone feared what they would find when the blood trail was followed up in the morning.

At the first light of dawn, twenty-five men carrying spears, axes and clubs gathered in front of Sukanti's home. The search party was led by the village leaders, Niladri Patro, Meena Rao and Gourhari. The group followed the faintly discernible dried drops of blood to a deep ravine a short distance to the west of the village. In the bottom of the ravine were two huge rounded rocks. The trail led to a small sheltered place behind the boulders and there, in a bloody jumble of gnawed flesh, lay the remains of the once vivacious girl, Sukanti. Only her left foot and head, with sightless eyes looking at her rescuers, remained. For several moments everyone was mute staring at the horrible sight before them. Then, one after the other, the men began casting furtive glances about them in case the killer might still be lurking nearby and watching them.

The visible evidence next to the remains gave few clues as to the killer's identity. Some men speculated that a huge tiger had carried off Sukanti. Others wondered about the other recent victims in adjoining villages where some evidence had pointed to a leopard as the killer. One man searching the ravine a few yards away called to the group saying that he had found a clear track visible in the sand. Quickly everyone gathered and examined the unmistakable pug mark of a huge male leopard. Sukanti's killer had been identified, but now a cold chill entered everyone's bones wondering who the next victim might be now that the man-eater was known to be in their area. The few gruesome remains of Sukanti were quickly gathered up in a cloth and the group trotted the short distance back to Dangasurada to break the news of their discovery to Sukanti's grief-stricken family.

Mandolo, as village headman, jumped on his bicycle and peddled to the Police Post at Muniguda to report the latest killing to the authorities.

At my home in Calcutta I had received a message in mid-January from the *Tehsildar* of Bisamcuttack telling me of the man-killings in his area, and requesting that I come to assist them by shooting the killer tiger or leopard that was creating such a state of terror in the villages of Hanamathapur-Chandrapur. It was 1958, and the Christmas and New Year holidays had just passed.

I considered my schedule and pondered making the trip to assist the villagers. I had hunted the area several times before and knew the jungles of the sector intimately. Since the letter was as much

Village women going for water or working their fields were often selected as victims by the man-eater.

a personal appeal as an official request, I started making plans to go as soon as possible. On January 15, the second message reached me. It was from Mandolo and was written in Oriya which I had to have translated.

The letter described Sukanti's death at Dangasurada and urgently begged for my help. Although I had already decided to go, the letter touched me deeply since I had met Sukanti's father on an earlier hunt in the area and had seen Sukanti around the village. Her bright eyes and glossy black hair conjured up in my mind how they now must have looked, matted in blood and torn to shreds by the claws and fangs of the man-eater. I am ordinarily not an overly sentimental man, but I felt Sukanti's untimely death should be avenged and time was of the essence. Others would certainly meet an equally horrible fate if the killer were not quickly destroyed. I called out to Khalifa, my house servant, to start getting our gear ready for the trip to Dangasurada. I added that he should exercise considerable flexibility in packing since I had no idea of how long we might be gone. Man-killers have a regrettable way of not following set schedules!

Early on January 19, Khalifa and I arrived at the principal village of Muniguda. From there we proceeded the several miles on a dirt track to the Reserve Bungalow, accompanied by a group of Dangasurada villagers who were marketing at Muniguda when we arrived. Someone had apparently preceded us because when we reached the bungalow at noon, Mandolo and the Dangasurada village secretary, Meena Rao, soon joined us after hearing about my arrival. Meena Rao carried his muzzleloader with him and everyone was in a high state of agitation as they told me the circumstances of the attack and the girl's death.

The *chowkidar* (guard), named Raghu, as well as the entire village delegation, implored me not to go anywhere without a weapon at hand for they were sure the killer would soon strike again since no victim had been taken in several days. My servant, Khalifa, listened intently. As the horrible deaths were described in intimate detail, he trembled noticeably and moved closer to the group lest he be the first new victim to be pulled away into the encroaching evening shadows. He felt certain the killer was watching him at that very moment!

116

As the sun began to lower in the western sky, the village elders indicated that in the interest of safety they had better quickly make the half-mile walk back to Dangasurada before darkness fell. Walking close together and making as much noise as possible, they made their departure leaving us alone at the bungalow.

Raghu had provided the servant with a *tangia* (axe) to protect himself. He himself also carried a *tangia* and, with this weapon constantly in hand, every domestic chore in the bungalow became a burden. I figured that if we were ever to get wood gathered, a fire started and an evening meal prepared I would have to volunteer to do guard duty or my two men, in their fear, would get nothing accomplished.

I asked them to lay aside their axes and I would stand watch with my rifle. This greatly helped their nervousness and they soon set about getting things unpacked and organized without spending most of their time staring apprehensively into the darkness and jumping at every shadow.

For greater security from attack, we decided to use the bath for a cooking area rather than the side room of the bungalow. To reach the side room required crossing an open verandah, which in the dark could easily invite an attack by the man-eater. As an additional precaution, I lit two carbide lamps and three kerosene lanterns that I had brought along. One carbide light was placed in front and the other at the rear of the bungalow giving a reasonable degree of light in the compound. I doubted if a cat could sneak up on us unseen. I placed myself in a chair from which I could see both entry doors to the bungalow. These we kept open to give me an unobstructed view in case the leopard was to pay us an unannounced visit. Both Khalifa and Raghu begged me to shut the doors, convinced that the leopard was about to momentarily pounce on them. Again, all work came to a halt until I finally shut the doors, faced either with that decision or going without my supper. The men now relaxed a bit instead of constantly running back and forth, reporting to one another if my sentry duty was providing them adequate assurances against instant attack from the leopard.

We established our sleeping area in one of the rooms, with both Khalifa and Raghu sleeping only a few feet from the foot of my cot. Any greater separation, I knew, would have created another scene of instant panic among my men. I deliberately left both windows open because they had solid iron bars, but I personally checked the bolt on the two entry doors to make certain that they were securely locked before retiring. I felt we were now in no imminent danger from the man-eater and, as my servants settled in to sleep, I lit a pipe before going to bed myself.

I had barely struck the match when we all heard the rasping, sawing sounds of a patrolling leopard close to the bungalow. Shaking with fear, Raghu whispered that we were all about to die since the man-eater had now found us and was about to make his attack. I tried to pacify him by pointing out that no man-eater would be so stupid as to announce his coming and that, in all likelihood, this was a perfectly normal leopard out hunting for natural prey. I knew from earlier hunts in the area that there were several leopard living there and I seriously doubted that this was Sukanti's killer. Seeing that I was not going to kill the leopard in spite of their begging, Raghu and Khalifa finally gave up and went to sleep. The sawing sounds soon stopped as our leopard moved on and we spent a quiet, uneventful night.

In the morning, I made a circle around the bungalow looking for the spoor of the leopard that had serenaded us only a few hours before. The track was not hard to find and I carefully measured the pug marks, noting any peculiarities in my notebook so I would be able to distinguish these tracks from those of any other leopard we might later encounter. I particularly wanted to see what the tracks of Sukanti's killer looked like.

I walked the short distance to Dangasurada and met with Mandolo and some other villagers who led me to the hut from which Sukanti had been snatched and then guided me to the two boulders where the remains of her body had been found. We searched carefully, but the wind and weather had obliterated any tracks of the leopard which the men indicated they had found when retrieving the remains.

Mandolo and Meena Rao pulled me aside, out of hearing of the others, and with great dismay asked why I had not shot the leopard outside of the bungalow the preceding night. Obviously Raghu and Khalifa had already shared their terror with the villagers. Both Mandolo and Meena Rao strongly felt that any leopard, under the circumstances, was a suspect and they felt that, in any case, no other male leopard would be patrolling the area. Since Sukanti had been attacked by a male leopard in the area, it seemed logical to them that *any* male in the area should be destroyed. I pointed out that on an earlier hunt a few years before, I had shot three males around this village in four days' time. They felt that was irrelevant and we had no meeting of the minds on the matter. To soothe the villagers I did agree, however, to kill the patrolling leopard if he should be so audacious as to growl around the bungalow again.

In the next several days, no further killings were reported to the police. On January 25, a steady light rain had fallen in the morning causing temperatures to drop sharply. By afternoon the rain had become a drizzle and, having nothing better to do, I put on a light rain jacket to provide some protection from the weather for me and my Mauser and decided to walk to the village. I shouted to Khalifa that I intended to visit with Mandolo to talk over what action might be taken in the absence of any new killings. We all realized that without a new death the hunt for the man-eater was considerably handicapped. As evening approached, I decided I wanted to get back to the bungalow since dark clouds were already again covering the sky and the track was extremely muddy and difficult to negotiate.

Walking at a rather slow pace, I was soon within 300 yards of the bungalow which was still hidden from view. For no explainable reason, I started getting that gut feeling that all was not well. I could see nothing moving about me, but I felt I was not alone. I held the Mauser firmly, examining each bush and boulder along the trail. Actually the jungle was conspicuous by its absence. There were many scattered bushes and small trees to my left and the area adjoining the trail on the right was also quite open for at least a half-mile before the jungle took over. I knew, however, how little cover a leopard required to hide itself and I judiciously examined every grass tussock, bush and rock before me as I proceeded toward the bungalow. Attempting to look in three directions to my front, while at the same time guarding my rear, is easier said than done, and I was immensely relieved when an alert langur monkey started giving its alarm calls from a huge pipal tree in front of me. I now knew that my sixth sense warning me of danger had been well-founded and that the langur could already see what was still invisible to me.

Soon the bungalow came into view and I noticed that the doors were tightly shut and that no sound came from within.

In front of the bungalow was an L-shaped grain bin next to the *pipal* tree from which my langur sentry kept up a steady alarm call of "*Khok, khok, khokkor.*"

I was startled by a sudden chorus of frightened shouts from Khalifa and Raghu who were also aware of the still unseen animal and desperately awaiting my return. Seeing me they shouted to warn me. At that very instant, a huge leopard bounded in great leaps across the compound, over the track and into a recently harvested paddy field. Taken by surprise, I tried a quick snap shot at the racing animal which was a complete miss. I chambered another round and fired again and, although I wasn't certain, I thought I saw the leopard lurch in its flight. It leaped high in the air, indicating a possible hit, before it disappeared from my sight.

Very carefully, I examined the ground along its readily visible tracks in the rain-softened ground and was delighted when I found blood drops. Prudence, however, called a halt to the search because it was getting quite dark and a drizzling rain was again falling.

Early the next morning, Meena Rao and another villager joined me to look for the wounded leopard. We had only gone a few yards from where I had given up the search the preceding evening when we located the leopard lying dead and stiff in a small hollow a short way up the heavily wooded hill. My shot had struck somewhat far back but, although gut shot, the bullet had worked its way forward tearing up much of the internal viscera. Still, in spite of this fearful wound, the animal had gone more than 50 yards and climbed 20 yards up a steep hill. We marveled at its vitality.

Sukanti's killer was an old male in good condition that gave no clues as to why it had become a man-eater.

The three of us carefully examined the huge male leopard which pegged out at 7 feet 3 inches. It was an old male that still possessed all his teeth and surprisingly showed no blemishes on his body which might have explained his man-killing tendencies. Since humans were only killed sporadically, it seemed to indicate that he still had the faculties necessary to kill normal game, and humans were only either opportunistic killings or a bizarre want of a change in diet.

I wondered momentarily if this might indeed not be Sukanti's killer, but Meena Rao and the other village Shikaris again tried to convince me that no other male leopard would have been hunting in the area. The villagers were fully convinced the dead leopard was the dreaded man-eater.

I remained in the area for another six weeks, until the end of March, but no further human kills were reported. As time passed, I, too, became reconciled to the fact that the man-eater had indeed been bagged and that Sukanti's death had been fully avenged.

LEOPARD ON THE ROOFTOP

by

Pat James Byrne

as told to

Capt. John H. Brandt

A man disappears on a rainy night; another a few days later. A leopard is shot which everyone assumes is the man-eater. Then a night vigil in a deserted hut almost turns into a disaster as the real man-eater returns to stalk the hunter.....

Jadu Manji tried desperately to keep the tattered umbrella over the head of his wife Rongo as she attempted to shield her two-month-old infant from the incessant rain. The little family huddled under the eves of a small rural bus stop near the village of Dharampur in Orissa state. The thatch provided little protection as the rain dripped down on them. It was late December and the night was cool. The soaking rain made everyone shiver as they crowded close together for warmth. No one else was at the stop, probably recognizing that travel on such a night had better be postponed until another day. The family had important business at Balliguda and both the man and his wife hoped the bus would come soon. At least inside the bus it would be dry.

The rain turned into a drizzle as the evening progressed and Jadu Manji handed his wife the umbrella saying he needed to make a quick stop behind the hut before the bus arrived. Rongo smiled and told him to come right back. There was no one around and it was deathly quiet at the edge of the small village. Jadu Manji said he would only be a second and walked off into the darkness.

After several moments Rongo called out telling her husband to hurry. No answer came and Rongo wondered why her husband had gone so far from the hut. She called again, louder this time, but still no reply came. By now Rongo was becoming a bit annoyed as she pulled her shawl tighter about her and the baby. After several more minutes she stood up and went to the corner of the hut and shouted into the pitch black darkness telling Jadu Manji to be quick since she could hear the bus coming in the distance as it slithered along the muddy road to the village.

Rongo approached the driver as the bus came to a halt saying she and her husband were passengers but her husband had left and had not returned. Hearing what she had said, several of the bus passengers made rude jokes about her missing husband, and everyone laughed loudly when someone suggested that maybe he had gone off for a drink. Several inconsiderate passengers added that she should return to her village to locate her errant husband and not hold up the bus departure.

As the driver switched on his headlights, he could see a white, unidentified object lying on the track some fifty yards in front of the bus. Among the passengers was a man more knowledgeable than the rest who had been listening with growing concern to

Rongo's tale. The man was a police officer and announcing his identity he ordered the bus driver to pull up to the white object to see what it was. Soon everyone could clearly see in the headlights that the object was a piece of white garment called a *chaddar*. There was a marked hush among the passengers as the police officer retrieved the item and all could clearly see that it was brightly stained with blotches of blood.

Calling Rongo to his side, he showed her the grisly discovery. She quickly identified it as the *chaddar* which her husband had been wearing when he disappeared. There was now no longer any question as to what had happened and several passengers came out to console Rongo who was screaming loudly in anguish and fear. It was a fear that gripped everyone and the police officer commanded all the passengers back into the bus for safety. Without a further word being said, everyone knew immediately that Jadu Manji had fallen victim to either a tiger or leopard and had been carried away in the darkness. There had been no sound. No cry for help. Death had been swiftly and skillfully executed. Knowing there was little further that could be accomplished, the police officer ordered the bus driver to proceed to the next station at Nuagam Post Office, taking the distraught wife and child along with them.

I was staying at a small bungalow at Nuagam at the time and, seeing the bus approach shortly after breakfast, I stepped off my verandah to see if my mail had been brought for me. It was still cold and wet outside and I carefully sidestepped or jumped over the many puddles of rain water still on the road as I approached the bus. A crowd of people had gathered and as they saw me a group split off running toward me. Everyone was shouting at once and I could make no sense of what they were saying. The police officer gave an authoritative command of silence and the group quickly shut up. Stepping forward, pulling the hysterical Rongo and her child with him, he proceeded to tell me what had happened at

Dharampur the preceding night. Although no one was absolutely certain, everyone strongly suspected that a man-eater had carried off the missing man.

Questioning the crying woman was unproductive as all she could say was that Jadu Manji had walked off into the night saying he would be right back and had never returned. She had heard or seen nothing until they had discovered his bloody piece of clothing in the roadway after the arrival of the bus. Placing Rongo in the care of some relatives in the village, I made immediate arrangements to return to the scene of the tragedy to see what I could discover regarding the mysterious disappearance.

Later that morning, upon the return of the mail bus, my tracker Budiya and I, accompanied by the headman of Nuagam, returned to Dharampur which we reached within a half-hour. In the bright sunshine, after the night's rain, the situation did not look nearly as ominous as it had in the dark of the night before. We started searching the sides of the road near where the bloodstained garment had been found.

Within a few minutes we located the pug marks of a large leopard in the muddy earth along the drain of the road embankment. We followed the tracks until we could see where the leopard had dug in when launching itself in the attack on its victim. Having made the kill, the leopard had dragged Jadu Manji along the side of the road. While carrying off its victim, the bloody outer garment, found by the bus passengers on the road the night before, had fallen off.

We were joined by a number of other village men eager to help find the body. The clearly visible drag mark, which soon diverged from the road bank, went up a small, wooded hill. Next to a large boulder which had provided some shelter, the leopard had laid down the body and commenced to eat. The grisly remains shocked many of the villagers who had never before seen a human kill made by a leopard. The man-eater had torn into the stomach cavity and had eaten a large portion of the lower abdomen as well as parts of the rib cage and a portion of the breast. We examined Jadu Manji's skull which quickly explained why there had been no outcry during the attack. Four large punctures on the back of the head, which had penetrated the cranium, indicated that the killer had struck his victim from behind and, with one bite to the head, had killed him. Thankfully, he died an instantaneous death and probably never knew what had befallen him.

I had planned to sit up over the corpse in the event the man-eater planned to return to consume what was left of his kill. However, shortly the police officials from Balliguda arrived at the scene guided by villagers. Since officialdom had preference, they removed the body, preventing any possibility of shooting the man-eater, even if he had returned. Unable to alter the situation, I returned to Nuagam village as well.

Several days went by with no further reported kills by the man-eater. It was during this time that two young men from Bodali village had gone off on a drinking party and were returning home from the Balliguda market. Since alcohol tends to exaggerate bravery as well as to dull common sense and good judgment, the two headed for home despite full knowledge of the man-eater being in the area. Upon reaching the banks of the Kalepin River, one of the drinking duet decided he was unable to proceed further and said he would lie down for a nap. Unable to carry his companion, the older man of the two decided to go on alone and planned to send someone back for his inebriated drinking partner as soon as he reached the village.

Upon reaching his home and explaining what had happened, the drunk was chastised for being out alone when he might well have become a victim of the man-eater. Several men soon left for the Kalepin River to bring the other man to the safety of the village. They carried axes and flaming torches to protect themselves and light their way in the dark jungle.

Reaching the spot on the river bank that had been identified to them, they searched and called but could find no one. In the light of the torches someone soon saw a glint of reflection from an axe next to an empty liquor bottle. No one said it out loud, but the realization of what might have happened entered several heads simultaneously. Spreading out, one of the men came across a blood-soaked *dhoti* which had belonged to the missing man. No further examination was required for the entire group to realize that the man-eater had claimed another victim. Knowing they could do nothing further to help, the two men ran as fast as possible to the village to alert everyone and make their gruesome discovery known. A larger group now quickly formed to return to the area to attempt to locate the body.

On the river bank the tracks of a leopard were soon found.

On the river bank, the tracks of a leopard were soon found. Lying nearby was a small, tin mirror case which was quickly identified as belonging to the missing man. Knowing that they could do little more in the dark and knowing the danger they themselves were in, the group returned to the village with the intention of resuming the search as soon as it was light enough to see.

I was still sound asleep that morning unaware of what had happened at Bodali village. A messenger pounded on my door, awakening me, and breathlessly informed me of what had happened to the drunken reveler the night before. Quickly dressing, I asked the messenger who was accompanied by some Bodali village men to return to the place of the kill, adding that I would join them as soon as I could get my gear together.

I had a bicycle for transportation and was shortly at the place where the road crossed the river. The men escorted me to where the mirror case and the *dhoti* cloth had been found. Strangely, we found tracks of both a tiger and a leopard which momentarily confused the search. Soon, however, I located the place where the death struggle had occurred. A large stain of blood still was visible and the tracks of the leopard were without question those of the killer. The tiger had only been an innocent traveler through the area either before or soon after the attack. My main problem at the time was, were these the tracks of the same leopard as the one who had killed Jadu Manji at the bus stop a few days before? It seemed most logical that it was, since it appeared unlikely that two man-eaters would be operating in the same locality at the same time.

While contemplating the scene of the attack, a langur started calling from the nearby hillside. Some of the village men went to investigate and soon I heard shouts of, "Come quickly, Sahib!" Rushing to the site, we pushed through the circle of men who were staring at what little was left of the unfortunate man. As in the other killing, the man-eater had eaten part of the stomach, chest and viscera.

There was much flesh still remaining on the body and the village headman and I decided to attempt to sit up over the kill hoping that the man-eater would return. We sent my tracker and the men from the village back to the road to wait for us saying we would only stay until sundown. I told them to make noise on leaving so the leopard would think the entire group had departed.

We selected a comfortable spot at the base of a large thorn bush some forty yards up the hill from the body. We had a clear field of vision and I felt the thorn bush would give us reasonable protection from an unwanted attack from the rear.

We only intended to stay as long as there was sufficient light to shoot since I had not brought a flashlight with me and also did not relish a nighttime sit-up on the ground with the man-eater about. About a half-hour before sundown we began to hear clearly audible noises coming from our left, quickly closing the distance between us. We were amazed the man-eater would be so reckless as to approach in this manner, apparently taking no precaution whatsoever. I had never before experienced such an audacious killer as this and gripped my rifle knowing that in a moment the killer would reach the dead man's body. Its approach grew louder and louder. Anytime a man-eater comes to a kill it is always a moment of extreme tension and excitement. Every nerve in my body was taut. Every sense was alert. Then the large, dark body of an animal came into view making directly for the kill. I slowly placed the gun to my shoulder. Then the adrenalin rapidly drained as I looked at a wild pig rather than the leopard!

Knowing that the villagers would have never accepted the abhorrent possibility of a pig desecrating the dead body, I had no choice but to kill the animal which fell dead at the first shot.

Within moments I heard the excited shouting of my tracker and the villagers as they scurried up the hill anticipating the death of the man-eater. I knew they would be disappointed but the circumstances were such that there was no other alternative. It was

The leopard, a large male, had numerous festering porcupine quills in its foot.

a quiet and dejected group that returned to the village that night carrying with it the remains of the latest victim for proper burial.

On January 11th, I was called from my home late in the afternoon by a villager on a bicycle imploring me to quickly follow him to the burial ground near Nuagam. He had been on his way to town and had seen a leopard brazenly exhuming a body in broad daylight. He insisted that if we hurried we might still find him there. I found all this a bit hard to believe but quickly slung my rifle over my shoulder and we pedaled as fast we could to the place where he had seen the leopard. We parked the bicycles and quietly approached the burial site concealed by a small hedge. Peering cautiously over the top, I was amazed to see the leopard still pawing the ground not more than fifty yards away. I sighted over the hedge and fired a shot. The leopard spun about and made one gigantic leap and was gone.

Hearing the shot, several men joined us; although we searched diligently, it soon became too dark to see and we did not find the leopard. Returning the next morning, we resumed the search again, going a bit further into the bush, looking hopefully for blood spoor which, until then, had eluded us. With no signs to guide us, we were amazed when, quite by accident, we stumbled across the body of the dead animal. Regretfully, it had been found before we arrived by a hyena which had managed to gnaw on the body and had consumed quite a bit of it. Although this was disappointing, the highlight of the discovery was that the leopard, which was a big male, had a large number of festering porcupine quills deeply imbedded in its left forefoot. The quills must have caused the animal great pain and would certainly have handicapped it in pursuit of normal prey. This, coupled with the fact that the leopard had been attempting to exhume a human body, left little doubt that this was the man-eater. So much for optimistic and hasty conclusions which would soon almost cost us our lives.

January 12th dawned as another gloomy, rainy day, but since we felt we now had effectively taken care of the man-eater problem, I was pleased when the mid-day bus brought in a visitor to break the monotony of a long, lonely day. The man coming to my bungalow proved to be the schoolmaster from a larger village called Gunjibada which lies about five miles up the road from Dharampur. The man, Surja Singh, informed me that a tiger had

Skins of six leopard shot in early 1953 by Byrne. The Sandrekia man-eater is the second from the left.

killed a milk cow belonging to an elderly widow living in a tiny forest hamlet called Sandrekia. He had been asked by the people of the village to see if he could request my help in killing the animal. Listening to the description, I felt that the matter of shooting a cattle-killing tiger would be quite routine and, not having any other commitments, I told him I would be pleased to help. I borrowed a 12-gauge shotgun and some heavy L.G. shells and, accompanied by my tracker Budiya, we caught the mid-day bus to Dharampur. After leaving the bus and before beginning our walk to the hamlet, I started to organize my gear and discovered that my 5-cell flashlight, upon which I was entirely dependent for nighttime shooting, had been left on the bus accidentally. There was little we could do except continue our walk toward Sandrekia, although I felt that now the situation totally favored the tiger rather than us. I had with me a rustic lantern with a headlight reflector that had been left behind by American troops serving in India during the war. It was far from adequate but, under the circumstances, the best we had available. Budiya and I decided to give it a try, although we were not very optimistic.

Following Surja Singh's directions, we walked along a swiftly flowing mountain stream and soon, after a strenuous climb, reached the tiny hamlet which consisted of all of three huts. Sandrekia was certainly no metropolis but, because of its size, it didn't take long to find where the cow had been pulled down. The kill, we were told, was lying under a small hut located on a hillside a short distance from the village. A frail woman holding a small infant to her bosom volunteered to show us to the place and then added that she was the owner of the dead cow. She was visibly moved as she told us that this was the last animal of three that she had owned and that now she would be thrown upon the mercy of relatives to support her since the tiger had now killed her last surviving animal.

The structure where the cow had been killed was a corral made of sturdy planks buried upright in the ground as a wall. On top of this was a bamboo frame covered with grass which served as a roof. I examined the dead cow and the surroundings and quickly determined that the killer was a leopard rather than a tiger. The woman begged me to rid them of the killer and stressed that it made little difference to her if the cow had died on the fangs of a leopard or a tiger. Knowing the desperate situation that existed at Sandrekia, I agreed to do what I could to shoot the beast.

Urging the villagers to return to their homes down the hill, Budiya and I dragged the cow outside of the enclosure; we then decided that in the absence of a suitable tree we would start our vigil on the fragile roof of the cow pen. This gave us little security since it was only about five feet high, but it was the best we could arrange. We placed a blind of straw in front of us to partially shield our movements and give me a chance to get my rifle in shooting position should the leopard return. It soon turned cold as the sun went down, and we pulled our jacket collars up to protect us from the breeze that now blew with considerable force down the hillside.

All was quiet and we heard no sounds for the first two hours after sundown. Quite suddenly our tranquility was shattered by a violent push from underneath the bamboo roofing upon which we were seated. The unexpected jolt was a surprise we hadn't expected and since we had heard or seen nothing, we moved our position a bit to see if we could peer into the cow pen to see what had happened. In doing so, a loose board fell from the roof into the interior of the pen. A blood curdling roar, followed by a deep snarl immediately below us, brought instant realization that the leopard had managed to sneak into the enclosure without alerting us and had tried, unsuccessfully, to reach us through the roof. It was a most unpleasant situation to be in, but it was quickly brought to resolution when the leopard made a bound over the outer wall of the corral and disappeared.

Budiya and I decided that our vantage point wasn't the safest place to sit in the dark since the leopard now knew our position and might make a more successful stalk the next time if he should try again. We took our headlamp and, walking very cautiously through the darkness, proceeded to a small hut nearby thinking that perhaps it would offer more security for us than the first.

Reaching the hut we went inside and Budiya lit a small fire. We found, to our dismay, that the door to the hut could not be properly closed and that only a fragile stick was propped against it to keep it from swinging open.

Budiya, who had been quite shaken by our recent visit by the leopard, commented that in view of the aggressive behavior of the animal he fully believed it was the man-killer, and that the leopard we had shot a few days earlier with the porcupine quills in its foot had not been the man-eater after all. I was inclined to agree since our recent nocturnal visitor was abnormally unafraid of people and seemed determined to add a new victim to its list.

While discussing this possibility, we heard a soft movement against the door. It sounded almost like someone knocking to request permission to enter. For a moment we thought perhaps some hillmen, having seen us, were seeking shelter for the night. I flashed on my headlamp and directed the beam toward the door. It suddenly became obvious that something was trying to enter and that the tapping was caused by the door being pushed inward against the small stick which we had propped against it. We both came to the chilling realization that our visitor was the leopard trying again to reach us!

Being unsuccessful in opening the door, the leopard moved around the hut examining it for other means of entry. We listened carefully and I motioned for Budiya to pick up a large boulder in the corner of the hut, which had been used for a fireplace, and place it securely against the door.

Budiya and I sat in complete silence attempting to interpret sounds from outside as to what the leopard was doing. Our reverie came to a quick end as we heard the leopard jump onto the roof of the hut. Quickly this was followed by dust showering down on us accompanied by scratching and tearing sounds as the leopard attempted to make a hole through which he could enter the hut. There was now no longer any question that this was the man-eater!

The leopard, as if contemplating his next move, paused for a moment and I took this opportunity to switch on the light thinking that his attack had progressed to the point that the light would in no way now deter the leopard from his intended purpose of killing us. Through the roof I could see the extended paw of the killer as he was enlarging the hole in the thatch. Budiya gave a yell and the leopard, momentarily startled, withdrew his paw and stuck his head through the opening to survey the situation and perhaps look over his intended meal.

The leopard was only a few feet above my head and I could have touched him with the barrel tip. Before this unsurpassed opportunity to kill the man-eater could fade, I pulled the trigger at point blank range. The leopard slumped backward from the velocity of the shot, but did not fall off the roof. There was no sound and we felt sure he was dead. Walking carefully outside we could see the spotted form lying on the roof of the hut. The shotgun blast, directly in his face, had been more than the man-eater could take.

Early the next morning, some village men helped carry the leopard back to Nuagam. There was jubilation in Sandrekia and Dharampur and everyone felt they could now resume a normal way of life again without the constant fear of imminent and horrible death.

While skinning this man-eater, I found that he also had been partially handicapped by having a number of porcupine quills imbedded in his forepaws. Walking must have been indeed painful for him. However, the decisive factor as to why he had turned into a man-eater was an old gunshot wound in his right shoulder. It had obviously crippled him to the point that he could no longer hunt properly and had turned to humans as his easiest source of food. There was no doubt that he was the dreaded man-eater and no more human deaths were reported in the area.

To my chagrin, I found that, since I had not killed the leopard actively over a human kill, under existing law I could not claim the government reward. The gratitude of the village was, however, more than adequate compensation.

SCARFACE: DEATH IN DARK WATERS

by

Capt. John H. Brandt

Canoes are overturned in quiet coastal waters by a killer crocodile whose last victim is a young soldier. The hunt, amid vows of vengeance, has an unexpected and explosive ending.....

The radio operator at Nakornsithamarat had received word of the impending storm but, since typhoons are a rare occurrence on the Malayan Peninsula, he had not taken the weather warning very seriously. He had shut down operations and gone home for the night without notifying anyone.

Shortly, the small village of Lamthalupuk, which sits on a narrow strip of land across the Ao Nakorn Bay from Nakornsithamarat, would be totally destroyed. First the roar of wind had come from the Gulf of Siam to the east and most of the houses and fishing boats had been swept away. When the eye of the storm passed over the village everyone felt, unfortunately incorrectly, that the calm indicated the storm was now past. Shortly, the reverse winds which had forced water into the bay returned with a vengeance and waves eight to ten feet high washed across the village to destroy what little was left. A few fortunate individuals had tied themselves fast to trees and escaped the death by drowning to which so many of their friends and neighbors had succumbed.

I had been sent to survey the damage to determine what assistance could be rendered. The storm had washed away any land approaches to the village and, accompanied by a number of Provincial officials, we proceeded by motorized sampan across the bay. The outward passage had been smooth but on our return a sharp wind had whipped up, churning the shallow water of Ao Nakorn into four and five-foot waves. The sampan only rode a few inches above the water and bow waves soon started to spill over the freeboard. We all bailed frantically as the sun began to set, but by then the coastal mangrove was within sight. Soon, in complete darkness the boatman found the appropriate entry channel, which was totally invisible to me, and proceeded up a dark lagoon, overhung with creepers and vines, on the way back to a landing on the mainland. At that point one of my associates said, "At least we're lucky that we're not in Surathani because if our boat had tipped over we would only have drowned, but up there we would have ended up in the stomach of old Scarface." Since I was on my way to Surathani the following day, I called him aside and asked who and what old Scarface was. He had a most fascinating and horrifying tale to relate.

From where we were, the next province to the north on the peninsula was Surathani, which could only be reached by a train from Nakornsithamarat to Ban Doan Station, a distance of some 100 miles. There were no roads on the east coast at that time and the train was the primary means of transport other than walking. One common exception was that wherever small waterways existed people traveled by poled canoes or sampans propelled by a rat-tailed engine on a long, flexible shaft. For a period of several months, sampans around Surathani had regularly been attacked and turned over by a monster crocodile. Over a half-dozen people were known to have been dragged under and killed. A number of others were missing and were also presumed to be the victims of the crocodile. Some empty sampans had been found adrift with no one aboard and most likely the same horrible fate had befallen these occupants as well. No one knew exactly how many had died because the provinces had very poor reporting systems for such matters and many deaths went unrecorded. The villagers were terrified and many were afraid to go to the market. Canoe travel

A huge, salt water crocodile glides through still waters where it is virtually invisible until a quick rush brings rapid death to its intended victim.

was the only way to get there and no one knew exactly where the crocodile lurked or where it would attack next.

Boats would often be totally unaware of the creature's presence until the entire craft was heaved into the air, spilling the occupants into the water.

Boats would often be totally unaware of the creature's presence until the entire craft was heaved into the air spilling the occupants into the water. In the fear of the moment, with everyone desperately splashing for shore, the dead were often not accounted for until a head count could be made. The croc would pull someone under so quickly that often there was no time to even scream for help. Not that any help would be available in any case!

The crocodile had been seen by survivors to have a large white mark on the forehead which had given it the name, "Scarface." Whether someone early in the croc's life had fired upon it at night while out hunting causing a large head wound that had healed, or whether the animal had been injured by the propeller of a motorboat was not known, but the prominent disfiguration made him distinctive from any other large crocodiles in the area.

The animal involved was the so-called salt water or ocean-going crocodile which, in spite of its name, lives quite normally in fresh or brackish coastal waters. It is distributed extensively throughout southeast Asia and is the largest and most dangerous of the saurians in the area. Capable of ocean travel, it is found north as far as the Micronesian Island of Palau where I had had my first encounter with the dreaded creature some years earlier.

While living in Micronesia, I had been at Koror when an unfortunate villager had been brought to the hospital with both his

In this idyllic setting, an unfortunate victim was attacked in the over-water outhouse by a salt water croc.

buttocks torn off. Hemorrhaging extensively, he was at the point of death when the physicians began working on him. His tragic tale caused muffled laughter among the staff but wasn't funny to him at all. He had made a nighttime visit to the *benjo* which is an outhouse on stilts perched over the tidal flats. The tide was in and as he squatted, extended over the outhouse seat, a large croc had passed under him. Salt water crocodiles consider man a part of their normal diet and this particular croc had surged out of the water and bitten the entire rear end off the screaming man who had desperately hung on to keep from being pulled into the water!

From Micronesia the animals extend into the Melanesian Islands, Indonesia and Australia. Among the largest of the crocodiles, there are recorded monsters of over twenty feet being killed in Borneo. In Singapore I observed a skull taken in the last century that, according to the measurements of the head alone, might have come from a crocodile over thirty feet in length! The enormity of a creature this size is frightening. Most full-grown males are in the eighteen to nineteen-foot class with only a few reaching twenty feet in length.

Salt water crocs have been hunted extensively for their valuable skins, with those of lengths in the eight to nine-foot range being the most valuable. Consequently, they have been seriously depleted over much of their former range and, in the early 1970's, hunting of salt water crocs was fully restricted. With passage of such laws, all international trade in this animal was prohibited.

Most recently, in 1987, a tourist from Colorado was pulled under and killed by a huge crocodile in Australia. Immediately prior to that in northern Australia, an enormous bull croc had taken on the terrifying habit of attacking the outboard motors on small fishing boats. A number had been attacked and swamped with the panicked fishermen struggling to shore awaiting any moment to be pulled under by the monster. As this report came to the attention of the authorities, a number of attempts were made to destroy this animal. It became known by the name "Sweetheart." Although many fishermen aged rapidly as a result of their attacks, it was thought curious that Sweetheart only attacked the outboards and tore them savagely from their position on the stern. The men splashing in the water had never been directly attacked. This was only a small point in Sweetheart's favor but most of the victims had been virtually scared to death from their experience, even if they had no tooth marks to show for it.

When Sweetheart was finally killed it was thought that the huge nineteen-foot male may have mistakenly interpreted the purr of the outboard engine as the growling threat from another encroaching male crocodile. Lying on the bottom of the stream, the silhouette of a boat may have looked to him like a gliding croc on the surface. His attacks had been focused on the sound of the motor; his efforts to protect his home, although commendable, were experiences that would cause nightmares for years to come for the unfortunate victims. Sweetheart is now on display at the Darwin Museum and it is easy to envision the horror of a person subjected to an attack by such a creature. The few surviving victims of the croc at Surathani had similar gruesome experiences.

I proceeded the following morning by train to Ban Don Station passing by the huge mountain called Khao Luang, the highest point on the peninsula with elevations over 6,000 feet. Upon arrival I immediately asked what the situation was concerning Scarface. All I had heard was quickly reaffirmed.

From Surathani north to the junction of Tha Chang no roads crossed the swampy delta of the Phum Duang River which emptied into the Gulf of Siam in Ao Ban Don Bay. Several tributaries, including the Yan River which connected from the north and the Tha Phi which flowed into the Phum Duang from the south, appeared to be the primary home areas of Scarface. Most of the attacks had taken place in the delta area and the adjoining mass of mangrove-lined lagoons.

Surathani Province is an area of vast mountainous jungle with no road connection at that time to the adjoining Province of

Chumporn to the north or to any road across the entire Isthmus to Takua Pha in the west. The last settlement which had a track cut to it was the settlement of Kirirat Nikhom, almost in the middle of the vast central jungle. Just to the east of the settlement, the Yan River joined with the Phum Duang. It was here that I planned to start looking for the giant crocodile, proceeding eastward to the delta.

That night in Surathani I made arrangements for two canoes and purchased three new four-cell-battery flashlights. These, I felt, would throw an adequate beam to show up a croc's eyes at night. I had my .300 Mag along to hopefully use on the croc if we were fortunate enough to find him.

Capt. Brandt and two assistants prepare to launch a light canoe on the junction of the Phum Duang and Ta-Phi Rivers to look for "Scarface."

It took most of the next day to get to Kirirat Nikhom where we launched the canoes planning to drift downstream along the shore of the slow-moving stream until we reached the mouth of the Tha Phi.

The night was clear and stars were everywhere. We allowed the current to move us along as silently as possible. Occasionally we would strike a submerged log or rock and with every bump my heart would skip a beat and I would think, "Oh, my God, this is it! We're on top of Scarface and didn't see him!" But each time the smooth stream carried us along again without mishap. We periodically switched on the lights and on several occasions saw the red reflections of crocodile eyes along the bank or among the dense vegetation which grew into the water. But as quickly as they were spotted the eyes disappeared and it was impossible to even identify whether the croc was a big one or a small one. At daybreak we made camp on the river bank and I slung a jungle hammock with mosquito netting on the sides and a nylon cover over the top

A jungle camp has few luxuries other than a rain and insect-proof hammock.

to ward off the frequent rain showers. By afternoon, we were again ready to make preparations to continue on downstream.

The second night was as uneventful as the first one and several times when we passed small canoes the occupants always indicated that further downstream was the most likely place to find Scarface. We passed the mouth of the Tha Phi early the next morning and moved into the maze of channels that form the mouth of the Phum Duang near Tha Chang. Canoe traffic was getting heavier as we approached the more settled localities. We planned to spend the third night moving about in the delta area.

Late that afternoon we heard a large motor craft approaching upstream on one of the channels. Upon seeing us, the craft veered our way and beached by our camp. Several government officials were on board and indicated they had been looking for us. Scarface had made an attack on a boat at midday not far from where we were camped.

On the launch was a very upset and grieving young soldier who, in the midst of his tears, regularly stood up and, vowing revenge, shouted torrents of violence into the watery wilderness. I asked one of the boatmen who he was and why he was so distressed. I was told that the latest victim that Scarface had pulled under that day had been the soldier's younger brother, and he was vowing to kill the croc or give his life in the effort.

A crocodile, being unable to chew, grabs hold of its prey and pulls it under, never letting go of its grasp. Since it cannot bite off pieces to swallow, it often pulls its victims to an underwater hideaway where the body can be secured among roots or vegetation until decomposition allows it to soften sufficiently for the crocodile to eat. In eating, the croc grabs a secure hold and then twists its body in the water until the mouthful comes free. In this way, arms and legs are easily wrenched from the dead body.

We were asked to accompany the group to a nearby backwater where the attack had occurred. The plan was to use long poles to prod the edges of the huge pool to see if the body could be dislodged—if it could be found. At least the soldier intended to give his younger brother a decent funeral ceremony.

Several hours went by and the sun was rapidly setting behind the trees. My boat was poling along an embankment about 100 yards from the launch. Everything was silent as death as the grim search continued. Suddenly, we heard loud shouts from the launch

Gliding by canoe along the banks of the Phum Duang River in search of "Scarface."

indicating that the body had been found. Decomposing gases had forced it to the surface once the prodding poles had jostled it loose. We turned to join the launch when suddenly a massive explosion rocked our boat. In the semi-darkness my first thought was that the launch had exploded, but I could see no fire. We hung onto our canoe to keep it from tipping over when the air was rent by another muffled explosion. This time we were close enough to see the huge geyser of water.

The young soldier had pulled the pin of two fragmentation grenades and had tossed them, at the moment of detonation, into the vegetation near where his brother's body had surfaced hoping that the croc was lurking nearby. This was a most unexpected event and I wondered then, as I have wondered since, why

Unable to chew, crocs must tear their victims to pieces by violent thrashing of the carcass.

shrapnel didn't penetrate the boat bottoms. The water must have slowed the fragments down, but the concussion was of stunning magnitude.

Quickly, flashlights played on the surface of the water and loud shouts came from the launch that the crocodile had surfaced. Keeping our lights on the floating croc we proceeded to within fifteen yards not knowing for sure what the animal would do. It was still very much alive. We found out later that the grenades had apparently broken some ribs and had severely crippled the beast. Since crocodiles are known to be able to sustain massive injuries and still survive, I, together with two armed men on the launch, fired a series of nine shots. The shooting was difficult in the bobbing boats since every time we aimed, someone would move the flashlight off to some other point of interest until someone shouted a command of, "Dammit, hold still!" and with the proper lighting some good shots were finally made.

A rope was thrown over Scarface and he was taken in tow to the village. In his stomach was found a large collection of beads, a bracelet and a set of keys. Surprisingly, he was not as large as we had anticipated measuring slightly over eleven feet in length. Although not a giant among crocodile, Scarface was as big as any I had ever seen, and his opened jaws looked large enough to drive a VW inside without crowding!

I returned home fully satisfied with our killing of Scarface and felt assured that now canoe travel would again be safe in this area. The event was not forgotten, but I was to meet Scarface again.

Many months had elapsed and in the northeastern town of Korat, hundreds of miles away, where I happened to be visiting, a small fair was being held. My servants came to tell me that the famous man-eater, "Scarface," was going to be on display. Surely, he was dead, and I couldn't imagine what advertisements my servants had seen.

I drove into Korat and pushed through the crowd gathered around a sign that proclaimed, "Scarface, killer of hundreds of people." Allowing for such exaggeration at carnivals, I hurried to see the animal. Apparently, some Chinese entrepreneurs had purchased the crocodile and quickly saw that they could make some money with the animal. Since no taxidermists existed in that part of the world, they had concocted a trough large enough to hold

Scarface and had submerged him in a mixture of formaldehyde and alcohol. Poor Scarface had shrunk horribly and his mouth was grotesquely propped open. The white scar on his forehead was still prominently visible. For a small fee, viewers were permitted to stick their hands or heads into the gaping maw of Scarface's mouth to get the imagined feel of being swallowed by a true man-eater. I couldn't help but feel a bit sorry for him in that his days were ended in so undignified a display.

No one who has ever had a boat turned over by a crocodile, or who has seen red eyes glide over the waters at night, or has even seen a formaldehyde-soaked rendition of a salt water crocodile is ever likely to soon forget the experience. It would quickly establish a firm conviction that if God would grant a choice in manner of dying, death by a man-eating croc would not be the one to choose.

THE DEVIL BOAR OF DOI VIENG

by

Gordon Young

A horribly wounded Lahu tribesman is brought to camp after being slashed by a killer boar which turns a gaur hunt into a near fatal ambush encounter with a murderous wild pig.....

There have been a number of fierce, wild boars in my life as a hunter. In fact, I've had more dangerous encounters with them than with any other potentially dangerous big game. I refer, of course, to *true* wild boar, species *Sus scrofa*, as found in Asia and counterparted in Europe.

Lahu mountain tribesmen, with whom I've done most of my hunting in Thailand, Burma and Laos, have a saying, "If the boar is just up-slope from you, and he's looking right at you from fifty yards, no hunter is fast enough with his gun when the boar has sounded off with his pre-charge 'Oommph!'"

Well, *most* hunters, anyway, especially if they haven't hunted boars.

The most imposing individuals of *Sus scrofa* that I've seen were among those I saw, and the few I took, in Korea. These were distinctly the "white-whiskered" monsters out of Siberia. Yet, take away the heavy hair coat and you have the same animal in a tropical version of this muscled bundle of dynamite. Give this tropical phenotype a reason and he's going to come after you without the snows to slow him. He simply gives off one loud "Oommph!" and he has arrived.

Forever in my memory will be one particularly unforgettable boar, encountered on the mountain of Doi Vieng Pha in northern Thailand. Like other boars that didn't hesitate to charge me, whether wounded or not, this one was no exception to the ultimate anger-provoking cause of having *rattan* palm thorns lodged in his snout. These thorns, upon close inspection, will show tiny, one-way barbs that, once embedded, work deeper into the skin and muscle tissues of animals. Wild pigs in *rattan* jungles root under the palms and often pick up troublesome thorns. The more infected and deeper the thorns penetrate, the angrier becomes the pig. In a way, it is like making a man-eater out of a leopard that picked up small porcupine quills, never having seen man or been wounded by him.

I shouldn't have been up on Doi Vieng that August of 1959. It was in the middle of the rains and miserably wet and dreary. It made me dream of an earlier hunting trip, in the delightful dry of February, far across the wide Muang Fang valley to the west toward the Burma border. I had been chasing gaur, a particularly excellent bull, and had shot a young pig just that morning, giving my small party plenty of meat. I saw one herd alone that had over sixty head of wild pigs and a magnificent herd boar that I simply watched—and passed up. I had mentioned to my companions just that morning that if I had wanted to hunt pigs, that's where I would have gone. But I was on Doi Vieng, hunting gaur again, and destined—though I didn't know it just then—to be drawn into a fracus with the most formidable wild boar I'd ever encountered in my life.

The rain drizzled down endlessly. Up near the top of Doi Vieng it was cold, fortunately too cool for leeches, although we didn't escape these troublesome pests in the valleys. We were camped in an abandoned thatch-roofed field hut held up by flimsy bamboo posts. A few years before, a family of hill people had scratched out an opium poppy field there, but second growth had obliterated most of the slash-and-burn cultivation. We had put considerable effort into repairing the old hut and it was with great difficulty that we managed to keep the campfire going under the hut. I stood on the narrow porch of the small hut with my Lahu companion, Javalo. He had been a large baby, born on "Pig Day," so he got the handle, "Big Pig." He had become a Lahu hunter of almost legendary repute and, at forty, he was far and away the most respected master hunter within the Mong Pulong clans of Lahu. Neither of us had anything good to say about the dismal weather. We tried to dismiss the thought.

"But it takes this time of year to find gaur up here again," Javalo was saying. "You have to follow the bamboo shoots. Run out of them lower down the slopes and they come back up to higher ranges where shoots still pop up here and there." He spoke in the vernacular which I'd learned from my childhood in Burma.

I didn't have a chance to continue our conversation. A long cry from a human voice, an ululating yodel from far up the slope, startled us. Javalo gazed up the steep slope that went up to the main ridgeline and hollered back with a few long whoops. The voice came back, desperately, and we barely made out Lahu words pleading for assistance. Whoever it was apparently knew we were there, perhaps could see our campfire's smoke. Javalo hollered back, assuring that help would be on the way. We looked at each other incredulously. It wasn't a place where we expected to run into any other human beings. It wasn't unheard of that bandits sometimes tried tactics like that in other places. But here, on Doi Vieng?

In a little while we had two of our more agile and younger members, Java-eh (Little Pig) and Jameh (Gets Lost), climbing rapidly up the trail. Fifteen minutes later, with the rain still drizzling and the haze still covering the mountains, we saw the shapes of three men far up on the slope, the lead man carrying a heavy burden. When they got nearer, we saw that Jameh carried a man piggyback hurrying as fast as he could down the slippery path. Java-eh behind him called out, "Man's hurt! Pig tusks got him!"

When they got to the hut, the stranger stepped up the ladder of the hut and panted out, "Ai-ee! That's my brother. Boar chewed him up. We heard you people came up here to hunt . . . saw the smoke yesterday. Maybe you've got powerful white man's kind of medicines!" He carried two ancient muzzleloading shotguns and unslung these to lean against the side of the hut. He looked exhausted, somehow having hung onto both guns while carrying his wounded brother. We saw that they were the more southern Lahus who predominated among Thailand's Lahu tribal people.

We stretched the groaning man onto a tarp on his belly and I gasped when I saw his wounds which were mainly on the back of his torso and legs. I groaned inwardly. It was infinitely more than someone like me with a little better than average first aid training should have had to face. Under those circumstances, it would have made a surgeon groan too.

We cut off the man's torn homespuns and I was introduced to what a wild boar is capable of doing to a man. The worst of the wounds were ripped furrows across his back, several of them longer than a foot. The skin over his right ribcage was in ribbons, exposing the ribs which might have also been broken. The muscles of his right thigh had been deeply cut, sliced as though clawed by a bear. I know that my face was uncontrollably contorted as I began an immediate attempt to clean debris that had collected in the wounds. I was amazed that most of the bleeding had apparently stopped.

It was only a year earlier when I had been confronted with an almost identical situation at Whuay Mi when I encountered another tribesman who had been frightfully mauled by an outraged bear. Consequently, when entering the jungle, I now carried a fairly complete medical kit that a Thai doctor friend at Chiangmai had helped me put together. The first thing I did was to find a clamp and place it on a severed vein. Then I proceeded, under impossible conditions, to go after the dozens of rips and tears on the man using every available item in my medical kit. It took me five hours with the assistance of my Lahu helpers. We learned to become jungle surgeons before we were through that session, then took up nursing occupations thereafter.

The man was in great pain, groaning constantly and fading away into what I was sure were his final stages of shock. I debated, reckoning that it was chancy under the circumstances of my patient's great weakness, then decided to ease his pains with an injection of morphine. To this day, I will never understand how he was able to bear through this emergency treatment and effort on our part to save him. The odds were tremendously against him. I had no anesthetics except ethyl chloride spray; and I used up my precious bottle of that. We ran out of sutures early in the effort, resorting to ordinary thread which we sterilized as best we could. The sutures we made were simply countless. It helped a great deal that I am something of a taxidermist with experience in sewing skins together.

The whole undertaking was beyond our remotest imagination or expectation that anything so difficult could be presented. I look at it and still wonder how we did what we did. But in the end, we managed to close and sterilize—to degrees entirely unacceptable in hospitals—all the wounds. After that, we prayed for mercy. The Lahus were more confident than I was that the man would live; I personally believed that we had but one chance in all hell to see the man alive by dawn. There was nothing else we could do; the nearest hospital was five days away. The only hope lay right there in that field hut and what we could do. I could only hope that a shot of penicillin—which I very fortunately carried—would do wonders.

The penicillin did do wonders. Together with my primitive-living patient's jungle-tough, indomitable spirit.

By a miracle, this Lahu man, Awneh (Mashed Rice) did make it through the night and was still alive the next morning. Months later, he made it all the way to complete recovery, fantastically scarred as he was. He had mostly relied on sheer grit. I don't know how the various treatments he got at home could have kept him from serious infection, but mountain people do know jungle medicines, obviously. It had still looked very doubtful to me five days after we first treated him and when we all helped to carry him back to his village perched high on the upper slope of a heavily forested mountain.

Yet, the miracle—as far as I'm concerned—occurred and he was able to join me on a hunting trip the next year. It was Awneh who showed me a great deal more about the back country of Doi Vieng than we'd been able to learn on many previous trips. Very tragically, his brother Khapa (Gets Stuck), who had carried him so far before he met us, was killed by a bear nine months later. These men were from the village of Whuay Mi (Valley of the Bear), both of them intrepid Lahu hunters, and the source of many tales and legends which remain among their people to this day.

While Awneh recovered, Khapa took Javalo, Java-eh and I the next morning to the place where the boar had done its fearsome tusking of his brother. It was noon on the next day after Awneh was brought in before we were able to get enough of a break in the rain to get out for a look to see what we could do about avenging Khapa's brother. Perhaps, as importantly, we all wanted to hunt again in that particular area of Doi Vieng and didn't need a threatening, angry boar to come at us out of nowhere.

A Lahu village, with pagan prayer flags flying, perches on a shoulder of Doi Vieng Mountain where the killer boar lived.

It was dubious justice. But Khapa insisted that we must carry it out even if none of us were too keen to look for the boar. I could sense that it was almost a religious fervancy on the part of Khapa to find the boar and kill him. "There are evil forest spirits that ride such an animal," he had said, his eyes wide and fearful. We understood well enough. It wasn't quite what the rest of us believed or didn't believe concerning forest spirits. I thought, smiling at my recollection, that it was almost exactly what a Lisu man, Eh Long, had said to me about a bull gaur on the Mae Salak that had killed his brother, "Evil spirits ride upon those horns."

Khapa took us to the top of Doi Vieng's main ridge, then down a long slope to the head of a stream where the broomgrass was thick. This area marked an ancient opium field as well, but that would have been some forty years before. I saw immediately that it was a dangerous sort of second growth in which to hunt wild boar; there were few trails and visibility was much too close. My better judgment told me right there and then to give up this stupidity for the sake of attempting a questionable execution of a boar that probably, in his own right, had every justification to attack a man encroaching upon his territory, in defense of his turf. I thought that he'd probably never harm another man if people were simply smart enough to detour his area and respect his temperament.

But no. We weren't that smart, any of us. Besides, we demanded to hunt on that boar's turf.

"Over there, by the small gooseberry tree," Khapa said, pointing across a shallow ravine. "Awneh probably startled the boar from just above him when he was tracking a *khui-zuh* (sambar) through the grass there. He couldn't tell me all the story, but I know his gun didn't fire, cap didn't even go off, but it was hit by the hammer, I saw that. The boar got him before he could change caps."

I pictured the man being knocked down by the boar's sudden rush, then hunched over, exposing his back to the boar as he came back at him. That was why he'd sustained so much injury on the backs of his thighs and torso. After the many strikes, Awneh had passed out and the boar finally lost interest. Khapa had worked his way over and risked being attacked himself in his quick move to rescue his brother.

We stood there for some time studying the terrain. I turned to Java-eh, our master tracker who "thinks like an animal." He'd been doing his own concentrated study and was ready with a good, calculated guess. "Over there is where he probably beds down," he said, and pointed to heavy grass near a low outcropping of rocks.

"That's where you'd bed down if you were the boar?"

"That's where." I knew that his sharp eyes had seen pig trails even from that distance. Java-eh was invariably right, especially about the habits of wild pigs, the hunting of which was one of his most important occupations.

We circled above and toward the rocks. Java-eh stopped to look at some old tracks. "No tracks of other pigs around here. One lone boar, big, big! Look," he said, pointing with his bare toe to tracks the size of a yearling calf. "Anybody can see that he's bedded here more than a year. He moves from here, goes over there and makes a new nest. Don't you see?"

No, I didn't quite see his logic. Java-eh looked at the rest of us as though we understood very little of the most basic things in tracking. His further whispered explanation confused me even more. Javalo nodded his head in complete agreement. Later, it was explained to me that a normal part of Java-eh's estimations in the jungles considered such obviously important factors as wind direction and *crawl direction*—however he could fathom and calculate that—of ticks! "Simple," he shrugged, "ticks move with the wind and rain direction, and pigs like to move *away* from them. Can't blame them."

Thus, Java-eh, in a very short time, was able to deduce for us just where the old boar would be bedded. And an old, ostracized-from-the-herd boar it would have to be, judging from the size of its tracks and its lone status.

We sat down then, planning the approach very carefully. There was every expectation the contact would be violent—and dangerous!

Awneh's misfortunes were very, very fresh in my memory. But it seemed unlikely that the boar would have much of a chance to get at us before we could get in our shots if if we skirted the tall grass from above. Java-eh had my 30/30 Winchester, Javalo had my 30/06 Winchester, I carried my old '06 Springfield, and our new friend Khapa had his muzzleloader. We were a formidable fighting force. We approached with the highest confidence in ourselves, and prepared to throw stones into the grass to kick up the boar.

To gain the low, rocky outcropping above the place where we thought the boar might be lying up, it was necessary first to push through some thorny brambles and dense bracken ferns. The ground was hard here and very slippery from a rain-soaked covering of mossy slime which, in the rains, clings to the ground over such places from constant rain flow. It was hard going and I found myself panting hard by the time we reached half-way up the other slope, making for the rocks. I came to an abrupt halt behind Java-eh as he stooped to scrutinize fresh tracks.

I was startled by the sudden change in the expression on his face. Java-eh looked on the verge of panic! He whispered to me, "Go back quickly, noiselessly!" and half-pushed me to get started. I didn't ask any questions and turned. The other two saw us and retreated at once, understanding tacitly that Java-eh knew what he was doing. When we got clear of the grass, he stopped. I asked him what his sudden panic was all about.

"Ha! We nearly walked into the boar! His tracks crossed the trail just there, so fresh I could smell him! That boar *knows* we are here . . . so he's up and around to *ambush* us! A pig like that, that's tasted man-blood will do just that. This is no place to be!" Java-eh said excitedly. We moved again, going for a large rock that I'd marked earlier as possibly a good place to jump on in case I needed it. When Java-eh gets startled and moves fast, there is every reason in the world to believe that a man had better move fast, that danger looms. The little man knows. Any Lahu knows!

The Lahus were making a lot better time than I was. I could see that they were almost in a panic. It was only a brief moment, but they simply faded away from me. I got the spirit of it all just then and wanted to call out, "Hey, wait for me!" I was beginning to feel very lonely and foolish.

I felt that I must pause. An ominous silence closed in around me. I assured myself that I had no reason to panic, that I'd moved far enough away from the possible danger and there was now some distance between me and the boar. I paused, then stepped slowly out into a sparser patch of grass and brackens, keeping an eye out behind me. When I'd gotten about fifteen feet into the more open spot, I paused and faced back in the direction we had retreated from. Off to my right, I could see that Java-eh had already reached the rock and beckoned me to hurry and join him there. I felt relieved that I had better visibility now. I couldn't imagine just then, or accept any such thought, that the boar would be stalking us. It just wasn't heard of. Of the hundreds of hunters' tales around campfires, such a thing as a boar following the hunter was never mentioned.

As I took slow, cautious steps toward the big rock, I heard a faint snort just to the left and above me in the tall grass. Then there was a sudden swishing of grass and a quick, choppy, inhaled-exhaled coughing sound as the boar launched into a crashing charge. In a flash I realized—and could believe!—that until then, he'd actually been stalking me almost noiselessly, exactly as Java-eh had feared. Some thirty yards above, he had probably followed a trail which only he knew so very well.

My main asset for having survived some close-calls on the hunt seems to be that I steady down rather quickly. It has amazed me more than anyone else. I had a poor field of fire and it didn't look good. I raised my old '06 hoping that I'd have time for an aim, slipping the safety catch automatically. Not forgetting that safety catch is something that became forever ingrained into my soul years before when a gaur bull almost got me. But I still say that Nature again had made a wonderful provision for me just then to steady me as it did—to be able to think and react! It was somewhat like knowing that you are inebriated, yet finding yourself amazingly steady and clear for a brief moment. With both eyes wide open and looking over my rifle barrel, I found I was telling myself, "Remember to *aim* . . . one shot . . . *aim!*"

I saw the big boar's head break from the grass, at most, ten yards away just then and still only partially visible. To my left I heard Java-eh yelling. He might have shot too except that from his perspective the boar was hidden by the grass.

Five yards and closing fast, I got off a shot. It wasn't difficult to hit him at that range. His momentum was going to carry him right into me. I leaped to my left, even as I saw that the strike of my bullet was moving the boar slightly to my right. I'd managed to catch him just above the snout, ducking his head into the ground and causing him to veer into a rolling crash past me not more than a few yards away. His scream was weirdly unlike a pig; it sounded more like a bear. My bullet had changed the structures in his nasal passages.

My Lahu friends cheered loudly from the rock. I was all nerves again by the time they reached me. It couldn't have happened. But the boar was still and the tangled and bloodied bracken ferns attested to the fact that he had passed by me.

Later, Java-eh ran his thumb over the jagged-edged tusks. "It has been broken before," he said. "These jagged ends explain the bad wounds on Awneh. You know why this one was so bad? Come, I'll show you." Java-eh ran his hand over the boar's bloody snout and showed me a number of black spots that covered the skin surface behind the flaring nose. "Now you see that this one was a boar out of the *rattan* groves," he exclaimed.

Khapa stepped up and said, "*Rattan* thorns hit every pig. This one had more than thorns. He had a devil in him . . . there was a devil on Doi Vieng that looked for him and found him. Now it goes to another pig, maybe a bear. Always, there'll be devils on Doi Vieng!"

It was a big, ugly and battle-scarred old boar. I estimated his weight at close to 650 pounds, give or take some. He didn't have the head I wanted to mount because the skin on his face was mostly scar tissue, his ears were ripped and split. But his meat smoked out well and we carried a lot of it back to the Lahu village.

The boar was an old, badly scarred tusker that weighed 650 pounds when he was brought down.

From far down the valley came a mournful call which quickly crescendoed into a brisk "*Pee-o-peet, pee-o-peet.*" The peacock pheasant's alarm had begun shortly after I fired my rifle, a typical reaction on Doi Vieng and other high mountains in Thailand. Javalo grinned at me and said, "Hear what the bird says? He speaks Lahu . . . '*That's enough! That's enough!*'"

THE KILLER BOAR OF BANGARUDA

by

Pat James Byrne

as told to

Capt. John H. Brandt

BRANDT

A murderous giant boar that has killed seven people returns after a mysterious absence to literally slash a village school teacher in half, and has the audacity to attack the very bullock cart carrying the hunter coming to kill it.....

My old Shikari friend, Mohini, and I had been packing and repacking, sorting and resorting our camp gear and hunting equipment for an extended stay in the far jungles of Orissa state in pursuit of the Koraput man-eating tiger. At least, at that stage, that was what our plans were.

Almost a week earlier, in mid-November of 1958, a telegram had arrived at my home saying briefly, "Collect permit, shoot maneater, letter follows." It had been sent by the District Officer of Koraput and the message implied urgency since the tiger had already killed several people.

After the cumbersome task of getting our gear to the train and finally getting underway, we allowed ourselves the first respite we had had in many a day. At eight o'clock the following morning we were at Bissamcuttack where I picked up my shooting permit and received a briefing on the depredations of the man-eater. It was suggested that since the tiger had recently been quite active in the vicinity of the village of Dangasurada, that it might be a good place to headquarter and set up our camp. We were further assured that once we reached the jump-off point at Kutragarh village there would be a number of men available to assist us in transporting ourselves and our equipment to Dangasurada. Up to this point all seemed to be moving along at a well-oiled pace and we had no premonition of how quickly the well-planned program schedule would be altered.

I sent my servant, Khalifa, out to arrange to hire a truck which could transport us as far as the Vansadhara River on the banks of which the village of Kutragarh was located. The bone-rattling ride was far from a comfortable jaunt but we all arrived safely. Unfortunately, there was no one there to meet us as we had been promised. The village of Dangasurada where we intended to camp was still over eleven miles away over a virtually non-existent forest track. The situation was further complicated by the fact that the Vansadhara River had to be forded at least three times before reaching the village. Without porters we were temporarily at a standstill and after a few minutes' deliberation we decided to shelter for the night in the far from luxurious Revenue Rest shed.

Early the following morning we walked around the village attempting to find a few men willing to accompany us and carry our gear. The negative responses were uniform, with everyone giving the same lament, "Sahib, it would please us to help you but we are afraid of the man-eater and cannot leave our families alone." No amount of coercion or bonus payments could uncover a volunteer. By now we were quite dejected and wondered what to do next. To go without our equipment was unthinkable—as well as unwise.

As we sat contemplating our misfortune, a cart pulled by two white bullocks pulled up a short distance from us. The driver, a tall, bearded man with a dingy white shawl, squatted next to the cart and lit a handrolled cigarette, cupping his hands in the manner of the hill people about the cigarette to warm them in the chill morning air. The sun shone brightly on him and after a few moments he seemed to thaw out, like a cold snake lying on a sun-warmed flat stone, and getting up slowly he walked toward us. He stood tall and erect before me and saluting me he said that his name was Shibo from the village of Dangasurada. With a tinge of contempt in his voice, he said that although the other men were frightened of the tiger he harbored no such fears, particularly now since the Sahib with the big guns would be accompanying him. He ended the short introductory speech by saying he would transport us free of charge to Dangasurada. We were elated and quickly loaded the cart lest he change his mind.

We placed Kahlifa up front with the cart driver and Mohini made a space for himself as best he could at the back of the cart and pulled the tarp up over him. I decided to carry a rifle and walk until we reached the first river crossing. With a groan from the bullocks and creak of the huge cartwheels we started our journey blissfully unaware of how quickly our adventure would begin.

In a short while we reached the first ford and since we had a long walk still ahead of us I thought it imprudent to soak my boots by wading across and jumped onto the back of the cart with Mohini. On the opposite bank I jumped off and resumed my position ahead of the cart carefully scanning the bushes and jungle, which closed in tightly along the track, just in case the Koraput man-eater might be out looking for an early breakfast.

Soon we arrived at the second river crossing and I again jumped on the back for a ride across the river. Mohini and I spoke for a few moments and the warmth of the sun soon mesmerized us both and I also laid back on the gear and quickly dozed off. Our awakening was unceremonious, dramatic, rude—and deadly!

My ill-timed sleep was jarred by loud shouts from both Shibo and Khalifa followed by a horrendous jolt as we felt ourselves, along with the bullocks, cart and gear, toppled into the drainage ditch. Something struck me on the head and I blacked out. When

The killer boar had often attacked carts, and people were afraid to go out to gather firewood, work their gardens or walk the roads.

I came to, probably only seconds later, I was tightly pinned and unable to move. I heard Mohini moving about and called to him to come pull the cart off me. He cut through the tarp and I was soon able to crawl free. I had sustained a bad cut under my chin but taking quick inventory of all my movable body parts I decided I was still ambulatory. Mohini also was rattled but unhurt. Shibo and Khalifa were nowhere in sight.

Mohini had been asleep with me and was as perplexed about what had happened as I was. The cart was in shambles lying on its side in the ditch. The unfortunate bullocks were in desperate shape suffering from numerous deep lacerations which were bleeding profusely. We quickly cut the bullocks loose but they appeared to be in shock and injured to the point that they could do little but raise their heads. I feared for their survival.

I asked Mohini what he thought had happened and he quickly replied, "It must have been an elephant that charged the cart." It sounded logical but I soon saw the clear slot marks in the road dust of a huge boar and pointing these out to Mohini he and I quickly agreed that the catastrophe had been caused by a gigantic wild hog which in India often reach enormous size and weight. With an aggressive personality and the physical strength to back it up, a wild boar can be one of nature's most dangerous creatures when it takes it upon itself to unleash its violence and propensity for killing and directs them toward man.

Pulling our gear from the cart, we loaded up the rifles and decided we had better quickly find our two missing men. We called out but received no answer. Apprehensively we walked carefully down the road and in about a hundred yards found a blood drenched *dhoti* which had been worn by the cart driver. There was a blood trail leading from it and within a few yards we could see the body of Shibo lying in the ditch by the roadside. The amount of blood on the body made it seem unlikely that Shibo was still alive, but feeling his pulse I felt a weak beat. He was hanging onto life but only with the weakest grasp! He had lost a great deal of blood and was in deep shock. We propped him up against a boulder and tried to stop the flow of blood as best we could.

Shibo was conscious and weakly said that the cart had been attacked by a large boar which had slashed the bullocks and then tipped the cart over on its side. He had jumped off and tried to escape but the boar was quicker and had attacked him as he ran. The boar had struck him from behind and had slashed upward across his left buttock cutting a horrifying gash across it. It had then made another strike across his legs and had ripped his thighs through to the bone. I bandaged the wounds and gave him a good shot of brandy but realized how inadequate my medical care was considering the severity of the wounds.

Leaving Mohini with the cart driver I grabbed my rifle and took off down the road shouting at intervals for Khalifa. Within another hundred yards I could see Khalifa perched safely on top of a huge boulder looking pale and very frightened. He had responded to none of my calls and I indeed was fortunate that I had seen him. He was frozen with fright and when I asked him why he had not answered me, he said that he had seen the boar attack Shibo and after seeing what the creature was capable of doing he had been too frightened to call out for fear the boar would hear him and climb up on the boulder and kill him as well. I told him that I would protect him and he watchfully slid down from the boulder and we were soon all back at the site of the attack. There was no further sign of the boar which had apparently melted back into the forest.

Since Shibo could not be moved and was desperately in need of medical help, it was decided that Mohini and Khalifa would remain with him while I proceeded to Dangasurada which was now about three miles ahead. I felt that within an hour I could easily reach the village.

The quick walk was uneventful and upon reaching the last crossing of the river I saw on the opposite bank a large group of men who were just getting ready to meet us at Kutragarh. I told them what had happened to us and within minutes we had two bullock carts and at least twenty men hastening down the road to help bring my men and gear to the village.

We soon reached Mohini and the others and decided that it would be best if we placed Shibo in one of the carts so he could be transported to the health center at Muniguda. We loaded the rest of the equipment and ourselves into the second cart.

Other men from the village carefully cut the badly wounded bullocks loose and helped them out of the drainage ditch. Even though trembling and unsteady from loss of blood, the bullocks were still able to walk and were carefully led along behind the cart.

Arriving back at Dangasurada we were shown a building where we could camp and the headman soon ordered tea to be prepared. Quickly we were surrounded by virtually every man of the village eager to hear our story and to share with us what they knew of the killer boar. It was only then that I learned how fortunate we had been in our unscheduled catastrophe.

The headman told us that the boar was an extremely dangerous animal and that it had already killed seven people in the area and had slashed a number of others. The killings and slashings, however, had all occurred almost two years earlier. Till now, everyone had placed the killer boar out of their minds. At the time, two huge boar had been killed near the village and everyone assumed that one of the two had been the killer. The actual killer boar, wherever it had been in the interim, had now returned and people were again afraid to work their fields, gather firewood or travel the roads. Compounded by fear of the man-eating tiger, it created a very tense and frightening atmosphere. Everyone assured me of how happy they were that Mohini and I were there to rid them of these terrifying creatures. Our tiger hunt had now become a pig hunt—for a killer boar!

I agreed with the villagers that the true killer had escaped the hunters' bullets earlier when the other two boar were shot and had quite likely avoided man for quite some time, or perhaps had wandered off to a more secluded part of the scrub jungle in the rocky hills where no one had come in contact with it. That it had returned, however, was now no longer in doubt! I considered the gravity of the situation and since I now had a personal vendetta with this particular boar I told the headman that killing the boar was the highest priority before others were killed or injured, but we would also, at the same time, make an effort to kill the tiger.

We moved our gear into the Dangasurada Revenue Bungalow which was a one-room affair about a half-mile from the main village. It was totally isolated and from the bungalow we could see no other building or habitation. It was as if we were in a jungle entirely to ourselves. The river flowed behind the bungalow which was our water source. The kitchen area was reached by an open verandah and the latrine was several yards behind the building. All in all, with a man-eating tiger stalking the area, it was far from an ideal—or safe place—to be setting up housekeeping.

That night we heard the sawing rasps of a hunting leopard and at daylight the following morning we could several huge mugger crocodiles basking on the river bank below us. The jungle and its abundance of game made it an ideal place for a sportshunter, but we had far grimmer plans in mind.

Purchasing some buffalo calves we staked these out at several ideal places in the jungle covering a considerable area. We walked over vast areas of jungle but could find no fresh signs of the tiger nor were our buffalo calves taken. After a few days I began to doubt the honesty of the villagers in telling us the man-killer was in their area and thought perhaps they had done this only to entice us to their village to offer an element of protection. Then on the fourth day we discovered the carcass of a half-eaten sambar doe and nearby we found the clear pug marks of a tiger. At least we now knew one tiger hunted the area, but was it the man-eater?

We built a sturdy *machan* over the carcass and Mohini and one villager elected to sit up over the kill in the hopes that the tiger would return. The *machan* was a good mile from the nearest road and was located at the base of a range of low hills. We had to work fast so the hunter could be on the platform by late afternoon. We departed, with much noise, in case the tiger was about hoping it would believe we had all left in a group.

I returned to the bungalow and after dinner cleaned and oiled my rifle before retiring. The stillness of the night was suddenly shattered by a loud sounding crash in the rear of the bungalow causing me to jump. The bungalow watchman and Khalifa both entered the room with expressions of puzzled anxiety. None of us could determine the cause of the noise. I held my hands up and placed my finger to my lips and both men stopped in mid-stride and remained still. The noise had come from the back of the building and I picked up my shotgun and loaded it with "BB" shot and quietly crossed the room and opened the back door. Stepping outside I almost bumped into an elephant in the dusk which was standing next to the verandah support. Just then there was another crash as the elephant yanked on the support causing a section of tile roofing to cave in. The elephant was so engrossed with his destructive bent that he never noticed my presence. I threw up the shotgun and fired the left barrel at the side of his head. He squealed, and whirling about, made for the river. I fired the left barrel into his rapidly departing rear. I sat guard for about thirty minutes after the elephant left just in case he should return. This time I kept a .470 Express with me but the elephant did not return.

The next morning I inventoried the damage and saw that in addition to tearing up the verandah the elephant had also taken objection to our latrine and had wrenched off the door which lay some twenty feet away in the bush. Walking around back I heard elephants trumpeting nearby and reaching the river bank I could see a whole herd of the creatures watering directly behind the bungalow.

Cautioning the men not to expose themselves and to refrain from going near the river to get water I returned to the bungalow to retrieve my rifle. I crawled up to the river bank and getting a good secure bench rest position for the rifle on a dirt clod, I selected a large rock as a target directly next to the watering elephants. I fired a left and a right into the rock and total pandemonium broke loose. There were far more elephant than I had realized and I hastily counted at least forty females and calves along with a few young bulls all thundering out of the river rushing for the safety of the forest. In a few seconds all was again silent. The shots had apparently had the desired effect and for the rest of our stay we were not disturbed further by elephants which now gave the bungalow and us a wide berth.

I watched as the men filled the water jars just in case an elephant or two had lagged behind. We soon all returned safely to the bungalow and the servants set about preparing breakfast.

I had just finished eating when two men loaded down with bundles of firewood rushed up to the verandah. They threw down their loads and came to me babbling out a horrifying tale that they and four others had been out collecting wood when the killer boar had attacked them. They had run and believed they were the only survivors. It had all happened so quickly that they had not taken time to fully determine the fate of their companions.

I grabbed my rifle and requested the men to show me where the attack had taken place. I was chagrined and angry when both men refused to go back with me. I could understand their fright but without a guide it would be difficult to go there quickly. They did, however, agree to give me explicit directions assuring me that it would be easy to locate. The attack had occurred in a small causeway on the main track leading to the village of Bangaruda. It was only about a mile from the bungalow where I was staying.

I walked carefully through the bush and soon arrived at the place which the two men had described to me. I searched the area carefully but could find no pig tracks, no blood, no bodies. It did not take long to arrive at the conclusion that I had been set up and that the two woodcutters had thought this would be a good prank to play. Unhappy and angry, I walked quickly to Dangasurada determined that the last laugh would be mine. I went to the home of the headman and explained what had happened. Listening carefully, he shouted out a stern command to some men seated in front of the hut and within moments my two woodcutters were dragged up to the headman's house. Sheepishly acknowledging that they had thought it would be funny to report an attack by the killer-boar, they were far from amused when the headman had them tied to a tree and administered a good lashing to the pair with a switch. I doubt if they would ever repeat such a prank without first giving it a lot of serious thought!

When I returned to the bungalow I saw that Mohini had already returned. He reported that the tiger had not come back to the kill but everyone had been quite tense because a herd of elephants had been feeding throughout the night all around the *machan.* Although none had come near them it was the anticipation of trouble that had made them welcome the first light of day. During the evening they had heard the two shots from my .470 when I had fired at the herd of elephant on the river and Mohini had wondered all night if I had perhaps already been lucky and had killed the man-eater. Unfortunately, such was not the case.

Mohini was worn out from his night-long vigil and promptly went to bed. Knowing that if I stayed at the bungalow I would distract him, I took my .300 Savage and thought I'd take a walk in the forest to see what I could find of interest. I had barely left the bungalow compound when I saw a crowd of men approaching from the direction of the village.

As they came close I could recognize some of the men as being from nearby Bangaruda. They rushed up to me to tell me that Bomala Appalaswamy, the village school teacher, had been attacked by the killer-boar while he bicycled along the track to

Muniguda. Some of the men in the group said they had only been a short distance away from Bomala when the attack occurred and they had witnessed the entire gruesome killing. The boar had charged out without warning from the scrub alongside the track and the poor school teacher, unable to get away, had been bowled over and the frenzied boar had then mauled him viciously making horrendous slashes on his body with its huge tusks. Bomala had bled to death before anyone could help him. The crowd volunteered to accompany me to the place since it was quite close by.

I stopped in the village of Bangaruda on the way to view the body of the latest victim and requested of the weeping widow, surrounded by her three bewildered children, if I could examine the corpse. She agreed and moved aside to be comforted by some other village women.

The extreme damage caused by the boar's tusks is difficult to comprehend. The strength of the animal in making its slashes had cut through to the ribs on both sides of the unfortunate man's chest. He had been literally sawn in half!

As we left the village and I gained more details of when the killing had taken place, it became apparent that this attack had occurred in or very near the same place where the boar had turned over our cart and slashed the cart driver. Feeling that they had fulfilled their duty in telling me where the killer should be, the entire crowd beat a quick retreat and left me alone on the road to continue my walk.

As I walked, the picture of the mutilated body of the school teacher, kept itself firmly imprinted in my mind and I started envisioning horrifying scenarios of what could happen if the boar burst out on me before I heard or saw it and didn't have time for a killing shot. I consoled myself that a boar never made a stealthy approach on its victim such as a tiger or leopard would. A boar, doughty warrior that he is, would most likely let out an emphatic grunt, click its teeth and make a head-on charge with all the vigor and murderous intent of a tank! At least this is how I thought it would go. I didn't have long to wait.

As I approached a thick clump of cactus growing from a rocky hillock I suddenly heard the expected "Whoomf" and crash as the boar leaped from the rock pile about twenty yards from me and made a determined charge directly at me. The speed of a charging boar leaves little time to contemplate the situation.

Flicking off the safety catch and bringing the rifle to my shoulder and pointing it in the right direction lost several microseconds and the boar had closed the distance between us to a perilous degree. With a prayer and a hope, I made a hurried sight picture on the boar and touched off the trigger. The round caught the pig in the head but its momentum carried it onward and by now it was practically on top of me. I worked another round into the chamber and touched off a second shot from a distance of only a few feet. The boar tumbled in its run and flipped over, landing almost in front of me. Every hunter, at one time or another, has a hair-raising experience but without question this particular charge brought me closer to death than any I had ever experienced before.

Leaving the boar where he lay, I turned to walk back to Bangaruda to get help in carrying the pig. Before I had gone far I was met by villagers who had heard my shots and were fully confident that I had slain the monster. There was no need to tell them that there had been moments not long before when I hadn't been so sure of the outcome.

Reaching Bangaruda, the entire village came out to see the killer-boar. I had Mohini take a picture of me seated on the animal and we then proceeded to cut the huge nine-inch-long tusks loose from the lower jaw. We distributed the meat to the villagers, but one woman, the wife of the school teacher, understandably declined her share.

The huge boar, which had slashed numerous victims and killed seven, brought on its own demise after virtually slicing its eighth victim in half.

We stayed for ten more days in pursuit of the Koraput man-eater but unfortunately made no further contact. The man-eater's depredations continued after we were forced to return to Calcutta, but at least we felt elation over the fact that the killer-boar of Bangaruda would no longer perpetuate further horrors on the villages of the area.

THE MAULER OF WHUAY MI

by

Gordon Young

A huge Asian black bear has mauled nine hill tribesmen, killing six of them. The frightened villagers request help in killing the monster, turning a routine hunt into a highly dangerous stalk for a man-killer.....

Lahu hunters have a saying, "Once struck by a green viper, one recoils at green vines." Most of us don't need a first viper bite or even a near miss! There's a vivid, unnerving experience in meeting any of the several members of southeast Asia's "Emerald" vipers. But at a small mountain village near the Burma border in northern Thailand I found the saying modified by local circumstances to, "Once threatened by bear, one recoils at shadows in the bushes!"

A spring just above this village flows into a larger stream a mile below with a fitting Thai name, Whuay Mi, "Bear Creek." It is said that the whole valley had at one time been over-populated with bears of fierce dispositions; and this was evidenced by a number of mauled and disabled hunters at various villages further down the valley through which a small river, the Mae Talope, meanders toward the north. Such grim reminders discouraged hunters from the lowlands from hunting in the area, even after the number of bears had dwindled considerably.

The Lahu village of Whuay Mi in northern Thailand which was Gordon Young's headquarters.

The rains were over and it was a bright November day as we got well into the Doi Vieng Pha mountains, an area I had been to only once before. There were gaur and sambar of exceptional size and numbers beyond the rocky crags, a day's hike south of the village. In those days, we hadn't known of the more convenient southern approaches, so we went via the Lahu village where we planned to spend the night. I counted on getting two Lahu hunters from the village, twin brothers we had met previously, to join us and share their extraordinary knowledge of the Doi Vieng Pha area.

In the afternoon, we reached the main fork of the Mae Talope where "Bear Creek" begins. The sweet, cold, creek water kept us there to enjoy a prolonged rest break after the long, hot walk upstream. Just as we began to get our packs back on, we heard agitated voices of men coming down the steep trail above us. The sounds suggested they were struggling with something as they moved along. Then we saw their burden—a man on a makeshift bamboo stretcher. They were all from the Lahu village and they recognized us at once from our previous meeting. Their chief, Ja-ti, was with them. He came over to me as soon as they had set the stretcher down near the stream, his round face sweating and troubled.

"It's a bad time. Our brother here is hurt, maybe beyond repair. Bear got him," he said, hunkering down on the trail next to me. "We go to the big city, Chiangmai—maybe their medicine people can heal him. Do you think they might be able to do that?" he asked.

From where I stood, I could see that the man was in a very serious condition. My medical kit, a gift from U.S. Army friends, was one of the items I carried for myself; it had been much improved, but suddenly it seemed very inadequate in the face of major injuries. Besides, what we needed at the moment was an expert surgeon, not an amateur medic! Then I felt a new worry grip me when a putrid odor wafted over the air.

Gangrene!

"Chief, it seems you've waited too long," I said, and knew that I looked as ill as I felt just then.

He ignored my comment. "The same bear took six other people. This one will be the first to live—if he lives." Chief Ja-ti looked at me then, surprising me with his sudden optimism, "But it seems we're guided by kind spirits today. We meet you here. I know you have strong medicines. You just fix him and we go back home tonight!"

I wished I could have shared his simple confidence in me. I looked at the bear-mauled mess of a man before me, holding my breath against the stench and unable to find anything to say to him. He looked fearfully at me from the corner of his remaining eye. The other one was a ghastly, blood-caked wound.

When they had pulled the filthy, homespun cloth off the man's torso, I choked and held back a powerful urge to vomit. I'm no doctor. I tried to brace myself, but I wasn't able to hide my horror. It seemed just then that the man didn't have a prayer to survive his ordeal for another hour, let alone make it to a hospital. It would take them most of the night to reach Chiangdao where, with some more luck, they might hope to find the police jeep available to help them reach Chiangmai. I got a grip on myself and leaned over the man to study the mass of dried blood covering his right eye and ear, or where they *had* been. The ear too had been slashed away, hanging on a shriveled shred on his neck. Gangrene had invaded his mangled right elbow and arm. How had this man survived this long? I couldn't fathom the man's tenacity to life, or his stoicism, or that I was actually seeing all this horror. At the same time, I felt infinite pity for the man and for the simple faith these people had in me and my small medical kit.

I spoke to the chief in language he might understand, "Chief, certain spirits have reached the shoulder and arm wounds, made them worse. It will take much work and medicines to fix these wounds. We don't carry the big medicines. When you reach Chiangmai, go to Suan-dawk, the government hospital. I will write words for you on paper. The chief doctor there is my friend." Chief Ja-ti nodded solemnly.

I opened my medical pouch and started to clean out the head wounds as best I could, then sprinkled sulfa powder over them.

There wasn't much I could do on the shoulder and upper arm of the man. I was certain that the arm and much of the shoulder would have to be removed—if he made it to the surgeon's operating room. I was convinced that my efforts were futile, but I had to do something. It was a choice between further agitating the man's precarious condition, and applying some measure of hopefully valuable first aid. There was no time to really clean the wounds; the big gamble against time was in the journey ahead. By luck, I had penicillin, a rarity in those days, which friends at the Chiangmai hospital had given me only a week before. I gave the man an injection; then with the help of Pa-suh, one of my companions, we covered the wounds as best we could with clean dressing material from my kit.

On a sheet of paper from my notebook, I wrote:

Dear Dr. Rubiat,
*   If by some miracle this poor man arrives alive, I wish you much luck! I gave him a shot of penicillin.*
*   See you,*
*   Gordon*

I gave the note to Chief Ja-ti and we watched them move off down the trail with urgent, rapid strides. We spoke of little else for the rest of the hike up the long ridge line and around the mountains to the Lahu village. It was dusk when my three Lahu companions and I reached the village. We were amazed to see the people standing around as though they had expected us to arrive just then. I never asked them but I believe they expected their own men to return instead, with the mauled man a corpse.

This latest mauling, by what had now become a notorious killer bear, changed my plans in midstream. Before we had the evening meal cooked, I had decided to see what I could do to find the bear, kill him and enjoy a good and happy relationship with the

Whuay Mi Lahus thereafter. This was important, if for no other reason than that these people lived on the edge of the best hunting areas then known to me in northern Thailand.

No one had followed the bear, the villagers explained. The man had been mauled six days before and they knew roughly where the bear would probably be. It was a big male, wounded a number of times; probably the same bear that had mauled a total of nine people, killing six of them. He was aggressive, they said, only when met in the jungles; he had not come into the villages or field huts. I thought of how honest and accurate accounts of such things are when related by mountain people. The bear would have attained exaggerated and fantastic proportions had he been the subject of valley people. According to the Lahus' more believable accounts, he was, in any case, as mean a bear as might be met anywhere; and he had the experience to be diabolically cunning.

The second headman, who was also the village shaman, told me over noisy sips of boiling-hot green tea about the last victim to be killed by the bear. The man had been an opium addict, he said, but a good hunter. He and his young son had been out hunting silver pheasants in a deeply forested valley. The bear surprised them, rearing up a few yards in front, and the man's reaction was to release an arrow from his cocked bird bow into the bear's chest. He tried to run away after his son, but was quickly caught by the bear and mauled into a broken mess within earshot of the boy. The boy, guessing that his father was surely dead, kept running until he reached home. As it was, the men of the village had not dared to go down to find the man's body until noon the next day.

It was a sleepless night. I had to struggle to stay awake despite cup after cup of strong mountain tea; but my visitors from Whuay Mi village simply couldn't get enough of their questions answered about "A-meh-li-kah." Then there was also the constant noise from under the chief's house: ponies, cattle, pigs and chickens; all, it seemed, afflicted with every manner of itching and nervousness.

With a great deal of effort, I managed to get off quite early the next morning. Pa-suh and Ja-meh, my two regular hunting companions at that time, accompanied me. Both of them were good trackers, but they were going after bear for the first time. The village people assured us that we would have no trouble picking up the bear's tracks: "Just follow the biggest tracks—you won't confuse them with ordinary bears. They're different; they're so big a man has to stop and marvel!" The bear was also known to frequent a particular field a few miles to the north where pumpkins were ripening. And this particular bear, they explained, had a special love for ripening pumpkins.

A small boy led us to the field, pointed to where the tracks had last been seen down near the stream below, then wasted no time turning around and sprinting for home. Minutes after we reached the edge of the field, we found the day-old tracks of an enormous Asian black bear. He measured noticeably larger than any I had ever seen, which is to say, the front pad was a good half-inch wider than my own palm with thumb alongside. I estimated the probably well-fattened animal to weigh in the vicinity of five-hundred pounds. This one was obviously in a class by himself. I didn't think he would be fussy about hiding out from people, and that he would probably fill himself on pumpkins, go a mile or so, and then bed down. My companions were agreed that we'd catch up to him in an hour at the most. I prepared my 30/06 Springfield with 220-grain Silvertip ammunition and reminded myself to be very much on my toes and ready for the bear the moment we started following him. It was a situation in which I had to do the tracking. I didn't want anyone within the 180-degrees ahead of me as we moved along. In fact, my two companions made a joke of how I didn't have enough money to pay them to track the bear, with me following behind! There was every reason to believe that the bear would come out fast and angry, from any direction, so I had joked back to my friends that they should watch behind themselves too!

As we moved up the other side of the small stream at the base of the field, I mulled over some thoughts concerning bears. It was a good way to stay alert. I had killed a number of bears and wasn't new to the game. Twice, irate sows with cubs had chased me into desperate struggles to escape when I had been a boy, mercifully abandoning their chase when they decided that I only wanted out. In any event, I am convinced that Asian black bear, with their long, shaggy fur and huge, white "v" on their chests, have much of the grizzly's temperament, while very likely being more agile in dense jungles or when climbing trees. They have a special cunning in setting up ambushes, and their sudden, surprising speed has fooled many experienced hunters. In dense jungles, a rushing bear makes a poor target, and they are tough to kill or turn from a charge. Lahu hunters had warned me many times never to shoot a bear on a slope *above me*; they roll and tumble down the hillside, suddenly coming alive as they reach the shooter. Above all, bears are not to be mistaken for the "dumb" brutes people think they are; they have excellent hearing and eyesight and their sense of smell is superb. As we tracked into the darkly-shadowed stand of big trees, I re-emphasized to myself, "Remember, they're cunning, CUNNING!"

It was easy tracking at first. The ground was soft and moist under the big trees, but after the bear passed this shaded area, his tracks became mere scuffs on the dry earth near the ridge line. He walked along every available log or big rock, often leaping off one onto another. This made the tracking much more difficult and we had to move slowly; I had to stop frequently to get my companions to help locate the lost tracks. In an hour we realized that he had no immediate plans to bed down, and by noon, five hours later, I began to wonder if he had any intention at all of staying in the general area. He seemed bent on going in a northerly direction, taking the ridges for more direct routes. It seemed to us he was distancing himself from the easy pickings of the pumpkin field. I wondered why he deliberately by-passed the likely spots to bed down, why he didn't keep closer to the field for easy, convenient access when he felt hungry and slept out again. It did not occur to any of us that he was quite purposefully giving us the runaround, and that we were, in fact, traveling in a wide circle. We were foolishly assuming too many things. The bear never *assumed*; he made his moves deliberately!

We had our lunch of cold rice and jerky venison on a rocky ridge top where it took some fifteen minutes to find the faint and subtle scuffs that told us where the bear had gone after he had paused there on a flat rock. We marked the spot where a bit of moss had been faintly mashed, ate our lunch, then resumed the tracking. Soon we were in low *mai lai* bamboos, a vining variety that has characteristics much like *rattan* and tangles into dense thickets four to six feet above the ground. At one place, the bamboos had collapsed into a solid barrier and, instead of skirting this obstacle,, the bear had pushed through close to the ground and left a space which I too had to push and burrow through. This alerted me to old ambush tricks I had heard about: like bandits placing a log across the road, bears were said to purposely create obstacles to trap the pursuer. So I was alert to this and had my rifle extended in front of me as I eased through under the bamboos.

The bear, fortunately, did not come rushing out at me. But I froze unexpectedly in an awkward position with my head close to the ground to face, what could be, a surer form of death. A small but very mature-looking cobra raised up and quickly expanded its small hood directly in front of me, a foot or so, it seemed to me, from my nose! I've always been keenly interested in herpetology and I've collected hundreds of specimens for museums. But this one allowed my enthusiasm to flee so quickly that I never did have an opportunity to collect or identify that particular individual specimen. Who knows, it might have been a very rare new species. But it was a situation a bit like looking down the barrel of a cocked and loaded pistol at the time and all I could think about was not to blink my eyes or else I'd have some very potently harmful catalysts introduced into a highly circulatory spot such as my nose! I also thought of how neuro-toxic venom—and this was certainly a species of cobra or elapid—might react when introduced so close to the brain. I looked with the greatest of discomfort at the small head which quivered very briefly to remind me of its determination.

Holding my breath, I backed up a few inches very carefully, making every effort not to give the slightest indication that I might dare to insult that small horror further. It somehow forgave me, but remained rigidly in place to show me who was boss. When I dared, I swung the muzzle of my rifle around and pressed its lethal little head into the soft earth. Then I got out my long-bladed hunting knife to render the snake harmless. I thought about that incident many times when discussions got around to hunting hazards and close calls; it was certainly as near death as I've ever been anywhere at any time.

By mid-afternoon, the sun had "softened" and there was a mild cooling in the mountain air. We didn't seem to be getting anywhere close to the bear and knew that soon the darkening shadows of late afternoon would stop our tracking. Among other things, that was exactly what the bear had in mind for us, I'm sure. After moving along a small ridge between two shallow stream cuts, the bear crossed the right fork, went up a steep bank, then traversed the slope's side for about fifty yards to cross over the head of a small landslide. He had gotten past the slide, then doubled back again to the middle of the slide. At this spot, his tracks seemed to vanish, and I spent moments looking for further clues. Then I discovered that he had leaped straight down the slide, clearing it neatly and leaving no signs in the loose debris. He landed on a large, smooth rock some fifteen feet below and leaped again from the rock over some low bushes to land beside the small stream. This leaping obviously loosened his bowels because he left a great pile of partially-digested pumpkins just past the stream. The pile looked to be at least a day old with small dung-roller beetles working busily on it.

Now the tracking was easy, but the going was bad. *Rattan* palms added to the matted tangles of vining bamboos and their long, hood-thorned trailers clawed and pulled at me as I tried to move silently after the bear's tracks. This area was criss-crossed by wild pig trails, and to follow him I had to assume a low stoop, just short of being on my hands and knees. A tougher type of trailing bamboo called *mai-mo* covered the area; and this is even heavier and denser than *mai lai*, tending to mat down closer to the ground. Underneath these canopies are tunnels which weave in and around the different clump masses. In very old stands, such as this one, the tunnels are tightly closed in, but usually opened into trails used by wild pig and sambar. When I paused at the edge of the bamboos, Pa-suh came up quietly behind me, tapped my shoulder and reminded me in a low whisper that it was against every reasonable thinking to follow a bad bear in this type of jungle. He didn't have to remind me! I knew well enough that the bear was sure to be close now and it wasn't a good place to be meeting him. The dim light under the bamboos made matters worse. Pa-suh was very young, but he entertained better sense than I did just then. It was sheer madness on my part to even consider following the bear further.

My better judgment screamed at me to give it up right there, but the devil was urging me to keep on going. I told Pa-suh to go back to where my other boy waited and remain there until I called them or returned to them. He gave me an apprehensive look and whispered back, "If you mix with the bear, there's no way we can help! If you die, we can't get your body!" It wasn't a very comforting thought to hear him say that or to watch him slip off back along the trail. I turned then, devoting most of my attention to finding the bear himself with only a cursory glance to see where he left tracks. If I moved with utmost silence and lucked out, I would find him bedded down and not rushing at me. He might appear anywhere, in front, to the flanks, or even behind me! I sweated despite the cooling air and realized that the humidity was still high. I cursed the pesky gnats hovering around my eyes and ears; I needed every advantage of sight and hearing. Still, I felt a crazy obsession to go through with the business just then, certain that I would never repeat such an unwise venture again. Yet it had to be done just this one time! I moved out at a squat, proceeding no more than a yard every few minutes, slow enough to make no noise and avoiding meticulously every sprig of bamboo that might spring as I brushed it. I talked some confidence back into myself as I considered how noisy bears are when they become riled up, with the teeth-clacking, and the snorting. In this kind of jungle he'd surely make a lot of sound and he'd be a big target. I was certain that I would see him before he could recover from his sleepy awakening and that I could take a careful, unhurried shot.

A small tree shrew scrambled about in the bamboos to my right, springing twigs of dead wood and causing me to whirl around to face the sound. It was a humiliating test of my nerves. I found myself too tense; a person made mistakes when too tense. Some shooters have, under such circumstances, taken a shot only to find the chamber was empty! To the shooter, *that* wasn't necessarily the most important thing. The important lesson was that he might have discovered that he had *flinched* his shot despite all the efforts he had put into perfecting his shooting skills. I had just managed the equivalent of a flinch when the little shrew unnerved me; and like a contrite shooter, I had to resolve not to flinch on the next shot. Inwardly, I thanked the little shrew for reminding me to calm down and be better prepared than I had been.

It took me at least ten minutes to reach a small, cleared spot, an ancient pig wallow, where the bear had bedded down. He had been there so recently that a musky, wet-dog smell lingered. I planted a knee firmly on the ground and waited, my rifle ready. I didn't want to move anymore than I had to now and scanned around me very carefully. The silence just then was like in the depths of some dark, deep cave. I knew then that the bear had been warned of my approach, probably even before I stopped to whisper to Pa-suh. It was unnerving to think that such a big animal could move off without being heard when we were actually very close to him. The flicker of a small wren and the tree shrew had drawn my attention immediately. I had, in fact, moved too slowly, allowing the almost stilled air to carry my scent to the bear; had I moved a bit faster, I might have kept abreast of the air's movement and surprised the bear. I knew then that I was really in greater trouble than I thought. The bear's reactions weren't as they were supposed to be; there should have been sound, a lot of sounds, because he should be showing his rage, chomping his teeth and snorting. Instead he had slipped away like a phantom. Why? To choose a good ambush spot in order to get an advantage on me, of course.

I remained there next to his bedding spot, one knee on the ground and my rifle raised. The air felt as though it would explode. In such a situation, I found myself focusing attention on my rifle, reassuring myself that it was a trusty, fine, old piece, that it had come through for me in tight situations before. But my breathing seemed too noisy, a giveaway, and kept surging much too violently to be easily controlled. I got lower to the ground in order to be able to see more of the area around me under the bamboos. It wasn't much. Ten well-shadowed yards between the heavy stands at best. Had I stood up, I couldn't have seen beyond the bamboo leaves before me.

Bamboo jungles are neutral. They undo the hunter, as well as the game, in many ways. For the hunter, bullets deflect off the brittle stalks to miss or register a poor hit; hidden stalks trip; and sprigs are easily set off into twanging, rattling noises to warn the game. For some, it had been a serious, even fatal, misadventure; but for all hunters who have been in bamboos, a certain twig often has earned his energetic curse. But it was my bear on this occasion that got into trouble with the same bamboos he had been so very good at moving around in thus far. It was a minute, subtle thing, but enough to force him to move before he had intended, and it spoiled his well-laid ambush. My big break came from a sound much softer than what the tree shrew had managed for me.

I have to guess here, but I think the bear, hidden in a hollow behind a mass of bamboo stalks, raised his head to peer at me, springing a small sprig. A small wren might have produced the same sound. It caught my attention at once and I turned slightly to my left to face it. The bear blended perfectly into the shadows and watched me from less than ten yards away. I could see the long, oblong mass the instant it began changing shape, not as I expected, but silently, like a rolling mass of lava, as it began to surge forward toward me. It was a baffling, awesome sight. I knew it was the bear, but couldn't believe it for a second or so. Then I was searching for the bear's head which became more visible, given away by the lighter color of its muzzle. I aimed at the middle of his mass, hoping for the best, and squeezed off 220 grains of Silvertip at him.

There are facets of luck to the well-placed hit, because I didn't have time to pinpoint my aim. Yet, I couldn't have missed the bear at five yards! What could have happened was, of course, a poor hit that didn't stop him. My bullet stopped him; it hit him just above his snout entering the soft nasal processes, through his palate and into the base of his neck just under the skull. The impact flipped the big brute and he stopped, facing off to my right. I put another round high on the side of his neck for good measure, then sat flat on the ground to vent my great relief. Minutes later I was aware that my Lahu companions were whistling quail calls to get my attention. There was no reason to play quails, so I simply called out to them to join me. I doubt that I was able to hide my elation or to keep from sounding too triumphant.

It was over. Lying on his back, four feet in the air, was the biggest bear I'd ever seen. I wished that I could have weighed him. He was surely in the maximum class for his species. We needed now to clear enough bamboo from around the bear's carcass to have room to skin him out. After my friends got to me, the first thing *Ja-meh* did was to inspect the bear's chest around the base of the "v." There was the small scab over a shallow puncture,

Gordon Young and the Lahu tracker, Pa-suh (New Frog), sit next to the giant Asian black bear that had mauled nine people and killed six others.

where his last victim had planted an arrow meant for a bird. Skinning the bear later showed that this had penetrated no more than an inch. Then we looked at the claws and saw the dried blood in the grooves under the hooks.

Daylight began to fade fast, so we skinned out the bear in haste. We still had a long hike ahead of us, up a steep mountain and along a long ridge, back to the Lahu village.

A year later, I visited Whuay Mi village again. It was, in a way, to observe a small celebration that had to do with the big bear. The people were appreciative that I had "punished" the bear, but I wasn't the guest of honor. We honored a Lahu man who had come back from the dead. It was a miracle indeed of human tenacity and surgical skills. We toasted, in the Lahu way by eating pork and rice, our good friend, Dr. Rubiat, who couldn't be present at the occasion.

DISHTOES: A KILLER BULL

by

Gordon Young

A gaur bull with a pathological hatred of man and the local nickname of "Dishtoes" has already gored and killed six hill tribesmen and possibly a seventh. During the hunt he tries desperately to add some more victims to his list.....

In the early 1950's, before opium-planting Lisu tribesmen in northern Thailand whacked away the great trees of Doi Sam-muan's eastern slopes, there were grand herds of wild cattle there. These surely ranked among the most impressive and proudest to be found anywhere in south and southeast Asia where gaur, the

A claim to notoriety exists in the areas hunted in northern Thailand in that opium is a primary cash crop, shown being harvested here by a hill tribesman.

world's largest surviving species of wild cattle, roamed. Just a very few years later and I would have missed the opportunity to hunt gaur in these northern mountains.

To be sure, mountain tribesmen had lived on most of the high ranges in northern Thailand since some two hundred years before. Yet, more ancient opium cultivation had been limited mainly to the higher ridges of this mountain and its radiating ranges, as it had in most other northern locales. This spotty and limited destruction of natural habitats had little effect on the numbers of wild cattle that remained. Some benefits were even derived as more grasslands formed to give more grazing areas where there had been poppy fields. As long as local hunters still carried traditional and primitive guns, the balance of gaur numbers was still on the

A Lisu tribesman displays the highly effective crossbow, shooting poisoned bolts with which village hunters brought down the largest game animal of the jungle.

plus side. Even the coarse thatchgrass which took over most of these old fields was good for grazing, especially when newly-sprouted after annual burns.

But "progress" was not to be held back. After the Japanese invasion between 1942 and 1945, a large number of old military rifles ended up in the hands of tribal hunters who had used crossbows or much less effective guns before that. Wild cattle now grazed in these open areas at increasing risk to themselves, for hunters didn't have to stalk up as close as they had to before with ancient muzzleloaders. This tipped the balance against the gaur and other big game and hastened the day for their rapid demise. It is the same sad story found in many places around the world where wars, the ravages of slash-and-burn agriculture, indifferent law enforcement, and encroaching "civilization" have enhanced the final disappearance of great natural habitats and the wildlife along with it.

For a season, a few great gaur bulls defied this onslaught and became legends. One of them, dubbed "Dishtoes" by Lisu and Lahu hunters, was a legend even at the tender age of four years. But this I didn't know until I was well into the hunt for this big and elusive bull. Dishtoes was still young, about six years old, when he made his last charge. In six years at least six men, and possibly a seventh, all barefooted, agile tribal hunters, both from the Lisu as well as Lahu tribes, died on his horns. I came close to being added to that list myself.

Dishtoes should have been slow on the occasion of that last charge: he carried something in the vicinity of about a kilogram of lead embedded within the tissues of his anatomy which had come from many attempts made by tribal hunters to kill him. They had used both Japanese military rifles as well as muzzleloading guns. He had been nearly down several times, just standing there and receiving anything and everything a muzzleloader could come up with while he waited to die in a cornered situation; this included hastily chopped up segments of a man's silver bracelet, bird shot, anything. Running out of shot or straight-ball military ammunition that had passed cleanly through the bull, more than one hunter had abandoned him and gone home for resupply. Dishtoes somehow recovered from these harrowing barrages and wasn't waiting when hunters returned for what they expected to be the final execution. Small wonder that he had developed a special hatred of hunters!

My introduction to Dishtoes began in early 1956. With my gang of six Lahu companions, I stopped to camp near a Karen village on the Mae Taeng River north of Chiangmai, Thailand. We were all new to the area, exploring new hunting areas for the sheer joy of it. Here, near the village of Muang Khawng, we met Bui-bui, one of the most intrepid Karen hunters I've ever had the pleasure of knowing. He had the compulsive desire to hunt just like the Lisu and Lahu who lived on higher elevations and he knew the approaches to the Sam Muan area like the palm of his hand. Not knowing any of the mountain people at the time, this knowledge was just what we needed to get acquainted with the area. In his own hunting, Bui-bui seldom saw the Lahu or Lisu himself so he knew very little of the background for the particular gaur bull that he simply came to know as a very aggressive one.

"This is no ordinary *kating* (gaur)," he had explained, "it is a very big bull, young and with horns still growing, but very big in the body. But follow him, and he'll be waiting for you! Then, 'Khoomph!' he is coming at you like a mountain falling apart! Ai-ya-ya-ya!"

"You've seen him and he has charged you?"

"Foolish question! Anyone who has seen this bull has left diarrhea on the jungle leaves! Twice, *nai*, twice it happened already to me. I've seen him across the valley more than once, but each time I was on the same side of the mountain as he was, he charged. You use a rich man's gun there, you don't have to worry," Bui-bui said, eyeing my old 30/06 Springfield converted military rifle with unconcealed envy.

"There are many *kating* up there?"

"Where I hunt, up on the Mae Salak stream, more than one herd come to the big lick up there, what the Thai call *poang*. You know *poang*?"

I nodded. These licks were mainly sulphur seepages that every manner of big game frequented almost by scheduled dates, seemingly by habit, yet more likely for instinctive reasons in order to suppress stomach parasites. Native hunters assume that animals become addicted to these mineral licks, as though they were some sort of narcotics. In addition to the sulphur, the licks may offer trace portions of calcium and other minerals that animals try to select.

"Elephant, gaur, deer," Bui-bui continued, "and even leopards and tigers that follow them. They come to the big *poang*. As much as I can see by tracks, there must be four, maybe five herds of *kating*, with six to eight in each herd."

I was all ears. I didn't approve of hunting as most native lowland hunters do by waiting at the mineral licks at night, but such places were a great way to find trophy-size animals for tracking first thing in the morning after they had visited a mineral lick.

"When do we go up into the Mae Salak?"

"When chickens crow in the morning," Bui-bui beamed.

That hadn't been far off on that first night we came to know Bui-bui. Our camp, just off in the trees near the noisy Karen village, had been a better one for sitting around the campfire and talking than for sleeping. Among other problems, there were frequent visits from curious village people who simply wanted to sit around and see what a pale-faced foreigner looked like. Very likely they hadn't seen one—certainly not the newer generation— since the days of British teak logging some twenty years before. I was glad when the first hint of dawn came.

We got off to an early start. Bui-bui was there exuberantly vociferous even before the chickens started singing. We were well up into the mountains from the Mae Taeng's valley by noon and stopped to fix a lunch. Clouds rolled in from the west over the higher Sam Muan mountains and with them a brief downpour that had us huddled under plastic sheets until it was over. An annoying drizzle continued as we trudged up a long slope.

Over to our left and deep in the long ravine of the Mae Salak stream, the steeply-falling stream's waterfalls roared impressively. We were headed up toward the headwaters, another good day's hard climbing. By mid-afternoon, we reached a small meadow that was Bui-bui's frequent camping spot and settled there for the night. The drizzle had stopped and the sky showed blue through gaps in the fleecy cumulus clouds above the grey mist. It seemed that we had climbed much higher into the mountains than the 4,000 feet of elevation at that point. One of the Lahus, Jacosuh, brought in a leaf monkey he had shot with my .22 rifle. It served to make up the delicately tasty meat portion of the wild ferns-and-watercress stew we had with our rice. With plenty of peppers, garlic and fresh ginger, of course!

Before we reached the cave above Bui-bui's *poang* the next morning, he knelt down by some tracks implanted in old elephant skid marks near a small creek. They made Bui-bui's eyes light up. "This is the bull! The one with worn tips to his front hooves. He travels alone, like always. See how the front of the hoof is shortened from climbing all those rocky slides and steep places."

I got down next to him and inspected the tracks of Dishtoes for the first time. We had no idea then that men had already died on his horns. Like Bui-bui, we were all fascinated with a bull that was as aggressive as Bui-bui had described. And as big! His dished hoof-tips didn't disguise the width of his tracks which were as broad across as that of a water buffalo, an animal of much smaller body proportions than its splayed-out hoof prints might suggest.

As a rule of thumb for gaur, I use my hand with fingers extended: seven fingers across the tracks is an average bull gaur; eight, as Dishtoes had, is a very big-framed bull. Just 200 yards down the slope from where we stood just then a Lisu man was to die a year later, the brother of a man called Eh Long, who in the end told me very much more about this formidable bull. Just then, it was the tracks that held my awed attention.

"Shot at him twice with this old gun," Bui-bui said, patting his ancient muzzleloader. "Each time he laughed at me! Didn't even seem to notice I put large lead balls into him. I was only a mosquito attacking him. With your gun, we would have butchered meat."

I smiled at him, recalling how he had run loving hands over my old rifle, one that looked very much less formidable next to the sort of caliber I probably should have carried at the time for such as the tough hides of gaur. In the end, I was to purchase and favor a .375 Magnum for such undertakings. But to Bui-bui, the 30/06 was an ultimate in life, the "too-expensive" gun, as he put it, one for which a man would gladly trade his wife, children, pigs, chickens and buffaloes. At the time, I wasn't complaining, nor would I ever apologize for my beloved 30/06, one I'd picked up at a swap shop in Pasadena, California. Before I acquired the .375, this same Springfield took a charging bull elephant and a number of other gaur, banting, bear, tiger and some very tough old boar. I have always had the highest regard for the various loads in 30/06 calibers as among the best all-round choices for hunting in jungle environments.

We settled into Bui-bui's cave camp and sat out the next deluge of rain which came almost on cue. It was as far as we were going because we'd found the gaur tracks well below where we thought we'd find him. It wasn't the greatest of camps, but it did save us a lot of work to keep dry. The tracks had been our first and main reason to stop and camp, but the rains got so nasty in the afternoon that we'd have stopped anyway. It was like dusk at five in the afternoon, and still the water poured down on us. There wouldn't be any hunting or even scouting probes that day.

Dawn was an entirely different story. It was a fresh, dripping dawn, full of earthly fragrances and the sounds of many birds. Muddied by the violent rains during the night, the stream flowed wildly over many large boulders, swollen to the height of the bushes on the opposite bank. From the mouth of the cave, I watched hyperactive chipmunks scurrying along wet branches, exploring and inspecting minute details in the bark. Over the top of stately bamboos, I could see a tall deadwood, its uppermost branch supporting a serpent eagle, feathers ruffled up and looking enormous, preening in the glow of the morning sunlight. Different troops of gibbon apes tried to outdo the others in a wild din of whooping their morning songs, a contest that went up and down the long Mae Salak ravine and gorges. I hoped that Thailand's budding program to preserve national forests would be in time to save such a land before slash-and-burners destroyed it.

Three hours later, Bui-bui, Javalo and I had climbed over the top of the ridge behind and above the camp. We came into a grass-covered flat near the head of a short fork of the Mae Salak and stood there, squinting against the sun at a dark mass in low bamboos. It moved. Bui-bui nodded up and down vigorously. I raised my rifle. The mass moved again, very suddenly and very quickly. There was a loud snort and heavy crashing.

In an instant, the big bull was gone. It was Dishtoes.

Without a word, the three of us made our way as fast as we could across the grass and quickly found where the bull had stood. The big bull had been in the process of backing up into a stand of tangled mosquito grass where he had good concealment for an ambush. He knew we had spotted him. His tracks now led off to our right, where he had rushed through bamboos and *rattan* patches around the base of where a ridge rose abruptly from the flat valley. By hurrying after him, we had a chance of seeing him across the valley, even if we couldn't catch up to him.

Dishtoes lined up for a charge downhill at me.

This first contact with Dishtoes, however, ended in near-disaster for at least one of us. He led us all the way, making us believe that he was bent on getting distance between himself and us in the typical fashion of spooked animals. He took us first across a small stream and while we traversed the steep bank on the other side, he had somehow maneuvered quickly and silently above us. From wherever he had been hidden, he came down upon us with a loud, blasting snort.

Javalo slid behind a tree.

Bui-bui dove headlong to the right of the charge and flattened to the ground.

I flopped sidewise into a thick stand of stinging nettles, moving with much less agility than my barefooted friends. I probably yelled when the nettles burned me, but I can't recall it. I was only aware that the bull bounded through us like a boulder rolling down the slope.

I was too flabbergasted to think, let alone consider shooting at the instant hulk that had materialized out of the thickets. We had all been so certain that the bull had moved off well beyond where he had actually waited for us. I had been assuming what I wasn't supposed to *assume*, that the bull, once spooked, would go a long way before he stopped again, just like most "normal" gaur are supposed to do—and usually do. Obviously, there wasn't any such thing as a hard and fast rule. More embarrassingly, we all had guns, poised to be used. For seasoned hunters, that's supposed to take only a few split seconds. Quick thinking and quick shooting are things I've credited myself for doing fairly well on a number of previous occasions.

This wasn't one of those occasions. Perhaps for understandable reasons.

Dishtoes did, however, keep on going after his charge this time. It was another of his tricks to have us believe that he'd be just around the next bend and he wouldn't be there. Make us believe one thing and do another. Perhaps this is the more typical habit of charge-wise gaur bulls. Miss a calculated charge, then MOVE! But once a bull has his victim down, then that becomes a different story.

After we dragged back to camp that night, exhausted from trying to catch up to Dishtoes, Bui-bui told us about a former hunting companion who had been with him in Burma on a gaur hunt. A wounded gaur had charged the man and just as he scampered up into the first branches of a tree, the gaur hooked the man's thigh, peeling him off the tree and tossing him to the ground.

"The bull kept on going, but turned quickly because he knew he had my friend on the ground. My friend was possibly knocked out and couldn't get up and escape when the bull returned. Possibly he died from that fall. Either that or he played dead. You know what the bull did? No, he didn't chop him up with his sharp hooves—that he could have done, yes. But gaur bulls don't paw at a downed man like a domestic bull might do. He just *licks* the man," Bui-bui said.

"He what?" I asked, smiling at what sounded like campfire lies.

"He *licks* a man," Bui-bui said, with knowing smugness. "He licks and he licks and he licks, because he likes the salty taste. Pretty soon the skin wears off and he seems to like the bloody taste even better. With my friend, the bull started licking his head. His scalp was gone when I found him that evening. He was dead. Hadn't been trampled or anything else like that. Just a big horn wound on his thigh and no scalp."

During the night I awoke with a loud groan from a dream in which a bull gaur had me cradled in his arms and was licking my face!

Our continued efforts to track him after we broke up our cave camp the next morning went on for two days. We still could not catch up with the gaur bull. At our third camp for that hunt, Javalo killed a young sambar stag and we ate venison instead of wild beef. Over the delicious steaks, we made plans for our next trip into the Mae Salak, more than ever aroused at the challenge that Dishtoes presented. The bull had proved to us in that brief encounter that Bui-bui's only fascination—really an obsession—had good basis.

I made four more trips into the Mae Salak after that. It would be almost two years to the day, and on my sixth trip to the same general area, before we would end our tracking of this elusive gaur. There had been twenty-two actual tracking days on his trail:

I followed Dishtoes for about 150 miles of estimated actual distance, saw him nine times, and was charged four times by the same animal. By this time, my own enthusiasm had developed into something akin to Bui-bui's obsession. Dishtoes was becoming a phantom and I was beginning to believe, like Bui-bui, that this particular bull had very specialized demon characteristics! During this interim, I had hunted and killed a number of other gaur bulls, and most of them had been far from easy pushovers. But this bull of Mae Salak was a different story.

By the 8th of November, 1958, we had made the trip to the Mae Salak often enough to know all of the shortcuts, even without Bui-bui's help. On this hunt, Bui-bui again accompanied me, together with his nephew Tamu, and my usual gang of Lahu men consisting of Javalo, Java-eh, Janu, Jacosuh and Teh-ku. We reached the *poang* below the cave camp early in the morning, just three days after leaving Chiangmai, ninety long, foot-traveled kilometers over the hills. As usual, one of the men asked me jokingly to describe my predictions for the hunt. I had stated, "We arrive at the *poang* at 7 o'clock. There we see the big bull drinking sulphur water. I kill him with one shot. We start skinning at 7:30." The response was the usual derisive laughter.

"Maybe we track for ten days this time," Javalo guessed.

"No, it all showed up in my dream last night," I insisted.

Java-eh, our unmatched tracker, went on to assure us that we would find the big bull because his own dreams were very auspicious by Lahu hunter standards: they concerned erotic experiences with women. Surely this was it this time. In addition, I now carried a little more insurance, a Winchester .375 Magnum. It can be imagined what this did to Bui-bui's enthusiasm for fondling a beautiful rifle.

We got to the overlook point above the *poang* very close to 7 o'clock as I had hoped. But the bull wasn't standing at the licks. Javalo and I left the others and stalked down to have a look at the tracks. In no time, we picked up the dish-toed prints which had become so familiar to us. The big bull had moved to the right of the lick's main flat and stood under a banyan tree to spill off a huge puddle of urine there before crossing the small stream. The foam was still on the bull's urine. We motioned to the rest of the men above us to remain there with our packs, that we had fresh tracks and were going off to follow the bull.

I checked my Model 70, released the safety and moved out ahead of Javalo on the easily-visible wet tracks across the stream. The heavy magnum felt awkward in my hands; I was much more accustomed to my 30/06 Springfield, but I felt a grand reassurance that I had plenty of gun this time and was glad to be hefting it as I pushed through the bamboos. The gaur bull had passed so recently, brushing the same bamboo sprigs along the well-used game trail, that the sweet scent of gaur oil was very easy to pick up. I remember looking at my watch and noting that it was 7:10 just then and it made me laugh inwardly. As though it was possible to schedule such unpredictable events as meeting a gaur bull! Especially one as notorious, by now, as this one had proven to be.

The bull was through with his drink at the *poang*. He was due now to look for some casual browsing before finding himself a bedding spot for the main part of the day. His tracks showed nothing to indicate nervous behavior or hurry. Nothing had spooked him. I felt confident that he'd be no more than fifty yards or so above us on a small clearing that we knew very well by this time, having passed through it many times. I paused after a short climb under the bamboos and visualized the big bull standing in the opening. Then the thought occurred to me that he wouldn't be there simply because I was too sure that he'd be there.

But he was there! Standing half-quartered away, where he'd stopped in his uphill climb. It was almost exactly as I had envisioned it! This was the first time I'd observed him in plain view. I'd only seen him as a passing blur or a fleeting glimpse the other times. I couldn't believe my luck! This was much more than the culmination of ten minutes of tracking from the *poang*; it was a dream come true for me.

But I didn't have him cold either, not even when I could dare to think that I had surprised him. He had probably heard us approaching and looked back over his right shoulder squarely at where we stood behind the screen of low-hanging bamboos. He wasn't sure about us, but he was definitely suspicious, a nature that had brought him this far against the high odds that some hunter would have surprised him long before this. I thought the wind had been in our favor going down-stream that early in the morning, but I could have been wrong.

I stepped out quickly from the bamboos with my rifle up and he saw me just then, just as I fired, going for his heart. I was a split-second slow and realized that he'd reacted instantly, lurching into a turn. I saw him jerk upwards, arching his back and knew I'd hit him well. But instead of moving away, he kept on turning until he was lined up for a charge downhill at me!

I had a new round in the rifle's chamber as he snorted loudly, a giant sneeze, before leaping down the slope with huge strides. Javalo's voice was choked, "Watch it! Take cover!"

Javalo was exactly opposite to me from a line to the bull's charge. We separated in opposite directions, knowing that the bull was going to go right between us or hit one of us. I had my rifle up, rather feebly, I must admit, as I realized that it was more important for me to move out of the bull's way than to try to shoot again just then. So, almost too late, I stepped aside as the bull crashed through between Javalo and me.

I remember thinking that I must wait until the bull was past the line from me to Javalo before shooting. I was half-sitting to the bull's right when I fired the heavy magnum from my waist, almost close enough to ram the barrel into his side! I saw the bullet flick hair and hide too high over the heart and too far back. Then he was past us and quickly below us on the gentle slope above the stream. He was still clearly visible under the bamboos.

I looked at my pants and saw flecks of blood splattered over the front. The blood was frothy, coming from his lungs, and he had flipped his mouth as he tossed his head in an effort to hook me with one of his horns. Then I saw that the bull hadn't gone any further than the flat above the stream where he had stopped abruptly. He had turned to face us again, looking uphill at us in what I imagined was another effort to line himself up for a charge. I still believe to this day that Dishtoes had every intention of giving it one more good try, though mortally wounded.

Dishtoes, a killer gaur that fatally gored a half-dozen tribesmen, was a young bull with a pathological hatred for man before he was brought down by Young.

He stood there raising his great head up and down, looking more tired with each effort to raise his head. The third of my 300-grain bullets took him in the center of his broad chest as he raised his head to expose it and this one coursed down high enough over the brisket to meet the low-hanging heart that is so easy to misjudge on gaurs because of their very high shoulder hump.

The long chase was over for the most unforgettable gaur in my life.

Some 3,000 pounds of all-lean, wild beef lay now where he had finally collapsed. I measured him later as he lay on his side—six and a half feet tall at the withers. The horns weren't meant for the body size this young gaur bull carried. Much older gaur bulls I have killed, perhaps by as much as ten years, never had this size even if their horns were greater in mass and length. Had he gone the full benefit of a natural life-span, I can imagine that his horns would have been tops in the record books.

About a year later, I was able to gather further details about this bull's past. A Lisu man, Eh Long, who lived at Doi Sam-muan's only Lisu village in those days, came to my home in Chiangmai one evening in January. He had some interesting things to say about the bull gaur we called Dishtoes. He could recount by name each of six hunters—and there might have been more, according to Eh Long—who had died on the horns of this bull in the course of just over three years. The first two men had been Lahu hunters followed by four Lisu men. All of them had been skilled hunters who made a last fatal mistake by underestimating the bull's cunning. The last victim had been his brother, Eh Sha-pa. And it was because of this that he'd come: to see the horns of the bull he had himself followed so far, been charged by often, and shot at so many times.

Even my intrepid friend Bui-bui had not known of these deaths although he was certain that we were hunting a killer bull.

I relived the experience in a new light listening to the Lisu hunter that evening in Chiangmai. Dishtoes, as it turned out, had really chalked up many more victories over man's powers of reasoning than the six definite kills he'd made on men. There had been dozens of such occasions; and I thought to myself as Eh Long spoke that I, too, had been outwitted many times. With his own brand of reasoning and planning, the bull had survived over four years of well-devised, hunters' plans to kill him. He had executed superb, evasive maneuvers and charged with incredible speed and ferocity into blazing guns whenever he deemed it the best course to take for the moment. In the end, and even while I followed him, his most relentless antagonists had been the two Lisu brothers.

Eh Long related to me that at first he and his brother had started out following the bull's tracks routinely, estimating that he was a young bull with tender meat. They had wondered why this particular young bull wasn't part of a herd instead of being a loner at that prime breeding age. They hadn't known that the bull had already killed a Lahu hunter named Jamaw a few months before.

"When the bull got our scent that first time we saw him," Eh Long said, "he just snorted loudly and charged away like he was frightened off—the usual thing to expect. We followed him running. Then suddenly the bull just wheeled in his tracks and reversed his charge. We were two brothers looking very stupid as we ran like fools into a charging bull *heh-nu*! But he caught us completely by surprise—we'd never seen a bull do this sudden turning. Always, it was to stop and move to one side, to waylay from tall grass."

The two brothers, experienced and used to making fast decisions in the jungles, dropped quickly to the ground just as the bull reached them. The bull leaped over them. As Eh Long pointed out, gaur will do this rather than trample the prostrated man in his path. Possibly, a charging bull thinks just then that the man may attempt to trip him, or that he is a log in his path. As the bull passed them, Eh Long emptied his old muzzleloader into its rump. He had added sadly at that point, "I, too, helped to make a devil out of the Mae Salak bull with the turned-up hoof-tips."

I must comment that by then I had also been jumped over by a charging gaur. I had also learned that hunter's lesson, and much more when it came to gaur. Yes, indeed, I will always maintain, just flatten out on the ground and let that old bull leap over you! It is the exhilarating experience every gaur hunter ought to have!

Following the bull again on a cautious run, the brothers soon came to an open ridge just at the edge of tall grass. The ominous silence spelled trouble again. They braced for another charge even in the seconds that the bull burst out of the tall grass to their left and came pounding down the slope at them. "Again we were fools," Eh Long continued, "twice in the same morning and with the same bull. My gun missed fire, and you must believe that I came very close to getting gored to death right there! Many people don't believe that I put my hand out in front of me, held that horn long enough to be *pushed* out and away from him as he charged and sent me flying! My brother jumped away off the ridge and was well out of the way. But it happened so fast! How could we allow ourselves to get so close to him before we smelled him? Eh Sha was cut on his heel by a sharp rock, so bad that we had to go home after that."

During the next three years, the brothers saw the bull again many times. They also shot at him many more times. One morning, they found and encountered him near the cave which had been our own camp more than once. By then they had, as Eh Long described it, become cocky, knowing just how to escape the bull's charge, almost routinely expecting it to happen to them. But on this occasion, his brother had been stopped by an unseen sapling that held him just long enough in the bull's path so that he was impaled through the chest by the bull's left horn.

It had devastated Eh Long. Like a madman, he'd followed the bull alone seeking a vengeance that was never to be his. He made no bones about having been quite provoked that I had robbed him of his very urgent mission.

I had gone to the Mae Salak that last time just at harvest time, when Eh Long had been too busy to go hunting. Besides, he had become very exhausted by then from following the bull for so long. I had smiled at the old hunter whose heavy homespun tribesman's clothes reeked of smoke from many campfires. I was reminded just then of a tiger I had also hunted so very ambitiously after it had become a man-killer in the same general area of the Mae Taeng valley. The tiger, in the end, had been shot, not by me but by a Lisu hunter who had used his last round of ammunition from an old Japanese Arisaka rifle to kill the beast. That had robbed me of "my" tiger too, just as I had robbed Eh Long of his justified revenge on the killer bull of the Mae Salak.

SHAITAN: THE MAN-EATING MUGGER OF KURSELA-KATARIA

by

Thomas F. Martin

as told to

Pat Byrne

Shaitan, the great mugger crocodile, basked in the sunshine with his huge jaws agape allowing crows to pick his teeth. His stomach was pleasingly full from his latest victim. As with so many of his meals in the past score of years, it had all been very easy for him. Shaitan's home was near Kursela, close to the railway bridge where he had, only a few days before, pulled under a fisherman named Bhusan Mondal. The unfortunate fisherman, who had provided him his latest meal, regularly placed his fish traps in a bend of the river where it swirled around the railroad bridge. One of the baskets had become tangled and Bhusan,

demonstrating poor judgment and carelessness, had jumped out of his boat to retrieve the fish basket. The crocodile had been stealthily watching and it had all been over in a swirl of foaming water as Shaitan grabbed the fisherman from beneath and with a few powerful flicks of its tail, the giant saurian had headed for the center of the stream. Bhusan, with the croc's teeth firmly imbedded in his stomach and back, screamed in pain and panic for someone to help him. Normally, such pleas would have been futile and any other fishermen nearby would have beaten a hasty retreat for shore to save their own lives. In this case, a crocodile hunter, Kenny Campbell, happened to be at the railway gangman's hut and, hearing the screams, ran outside just in time to see the churning water with Bhusan in its hub. He quickly fired two rounds from a .270 close to the croc in hopes it would release its victim. Shaitan, however, had been shot at many times before and, securing a firmer grasp, he dove with his struggling victim who was never seen again.

"Shaitan" translates in Hindi as "the devil," and the killer croc had earned his name well. The people along the river between Kursela and Kataria rail stations had lived in dread of the monster for years. Many had tried to kill him but he seemed to possess magical powers that made him seem mystical in the eyes of the river people. It was 1935, and I was about to make my second attempt to kill the monster.....

The huge crocodiles of India (*Crocodylus gangeticus*), sometimes referred to as the Ganges crocodile, are, in India, most commonly called muggers, from the Hindi word *magar*. Muggers are flesh eaters, normally feeding on animals—often domestic cattle coming to the river to drink. Turtles and fish are also consumed, but the normal method of hunting is to lie stealthily in wait for a suitable victim to approach the river bank and then make a lightning snap with its jaws and pull its unfortunate prey into the water. Not all muggers become man-eaters, but once a predilection for human flesh is acquired, it becomes a scourge to river folk who must fish the waters and go for washing and drinking to the very place where the croc is most likely to lurk. Crocs, whether in Africa or India, kill more humans annually than any of the other great predators, and Shaitan had killed enough to acquire local notoriety and a name of his own.

Most major rivers of Bihar contained at that time large numbers of crocodiles, and many small insignificant backwaters and jungle pools, far removed from human habitation, were veritable havens for crocs. They could often be counted in the hundreds basking on suitable river embankments. Among the muggers were interspersed giant gavials, a long-snouted croc with a peculiar knob on its nose, which ate only fish. Although frightening in appearance, it was not in a class with the muggers as it pertained to man-killing. Everyone wanted the crocs destroyed and no permits from the government were required to shoot them. My friend, Cecil Pritchard, and I had our first encounter with Shaitan while we were both still novice hunters.

The rail stations of Kargola Road, Kursela and Kataria lay along the Bengal and North Western rail lines, and the river which followed the tracks was full of large crocodiles. A particularly popular place for shooting crocs was at a railway bridge between Kursela and Kataria. A huge lagoon had been created at the bridge where construction workers had excavated a deep pool in order to build up the river embankment to act as a spur to deflect the river currents. The pool was approximately 500 yards long by 150 yards wide. It was some twelve feet deep and formed an "L" at the point running parallel to the track on the approach to the bridge. The embankment was a favorite basking place for the crocs and as many as sixty had been counted in the area ranging normally in size from five to twelve feet.

Cecil and I had attempted, unsuccessfully, to get a shot at Shaitan in the area; but, although we had experienced some good

219

stalks, Shaitan was always too alert to permit us a killing shot. In the winter of 1934-35, we decided to take off a few weekends to try again between Kursela and Kataria where Shaitan was regularly killing people with dreaded frequency. Every village along the river had tales of horror to relate to us and begged us to do something to remove the menace. Fishermen were afraid to go out, cowherds feared bringing their stock to water, and village women hesitated to wash and draw water for fear of losing their lives. Try as we would, with painstaking care, Shaitan was always smarter and the season passed with no luck. We had to wait until late fall before time permitted us to again return in pursuit of the man-eater.

It was mid-December of 1935 when we boarded the Sonepur passenger train at Katihar at midnight for the three-hour run to Kataria. Since it was still several hours until daylight when we arrived, we decided to walk to the hut of a headman, Souren Mahsi, whom we knew at a small fishing hamlet on the nearby Sunkosi River. We arrived as Souren was awakening and told him how we were again planning to hunt Shaitan. The eyes of the old man brightened as he stoked his fire and began telling us that he personally knew of Shaitan for almost thirty years and that all efforts so far to kill the creature had been futile. Many professional Shikaris had tried, but all had been uniformly unsuccessful. Souren added that he felt confident that much of the problem was due to the fact that over the years Shaitan had come to recognize the difference between the scantily-clad villagers wearing *dhotis* or *saris* and what he perceived as a hunter, who might harm him, dressed in khaki long trousers, khaki shirt and topped off with a solar *topee* hat on his head. With this classical shikar uniform before him, Shaitan remained illusively hidden, recognizing the danger to himself. He had been shot at and wounded a number of times over the years. Unfortunately, he had always survived but had gotten just a bit smarter and more cautious with each new bullet hole in his hide. Souren finished his sage counsel to us by saying that a young village girl had been taken by the croc only ten days earlier. She had gone to the river bank and had never been seen again. There was no doubt in anyone's mind that Shaitan had secured another meal.

Cecil and I judged that, in view of the recent kill, Shaitan might well still be in the vicinity of the village, perhaps lying low awaiting another easy meal. We took leave of Souren and made our way almost a half-mile upstream where we hoped to conceal ourselves behind the undulating edges of the river bank. We had brought along some old S & W fruit cans from which we had removed both tops and bottoms. We made little notches in the embankment where we installed these so that we might be able to observe the river below us without risking the danger of silhouetting ourselves above the embankment and frightening our quarry off. The tins would also provide space as gun ports through which we might shoot, if Shaitan showed himself, without having to maneuver into a shooting position and possibly exposing ourselves to the alert croc.

As the low-hanging fog dissipated about 11 a.m., we both watched, fascinated, as a huge crocodile crawled with great deliberation onto a sandbar on the river's edge only a short distance from our observation post. It was an extremely large animal and we wondered, hopefully, if it was indeed Shaitan. No such opportunity had presented itself before on our several earlier unsuccessful hunts. We carefully estimated the distance at about 200 yards. We had a pair of 8x24 binoculars and could observe the basking crocodile quite well, but we both agreed that attempting to make a shot at that distance, with open sights, would only add another wound to the killer and allowed us little hope of making an anchoring shot that would keep him in place until we could finish him off. Whether he was Shaitan or not, he was indeed a splendid trophy and we did not want him to get away.

We agreed that I would remain in place and Cecil would make the stalk, behind the embankment, to get as close as possible. Cecil crouched on hands and knees and slithered forward to make his body as small as possible and to reduce his silhouette against the white sand. Cecil crawled for some one hundred yards before taking his first rest. He looked back at me and I signalled that all still appeared well. I could clearly see the croc, who now appeared to be fast asleep. His jaws were open and several crows were picking his teeth, oblivious to the danger if the huge jaws should suddenly snap shut. The mugger was lying diagonally across my line of sight, but to the position where Cecil had now crawled, the croc faced him virtually head-on. I looked at Cecil and saw that he had begun his second crawl which appeared to be taking him to a point not more than 75 yards from the sleeping monster. Minutes ticked by until Cecil very, very slowly raised the rifle to his shoulder and making a minimum of movement, sighted and released the first round from the .318 Express. The shot seemed to lift the entire front end of the mugger croc who let out a horrifying bellow and attempted a cumbersome slow-motion move to reach the water. Cecil touched off the second round and with a few feeble flicks of its tail, the huge saurian shuddered and lay still. Neither Cecil nor I moved from our position.

We let several minutes go by to be certain the mugger was indeed dead. I moved up to join Cecil and from a distance of 20 feet we kept our rifles trained on the creature just in the remote event that its death was being pantomimed. Soon though, there was no further question that the croc was dead. On hearing the shots, many villagers ran over to see what had happened. Souren arrived with two long sturdy poles which he used to navigate his boat on the river. He prodded the mugger a few times to reaffirm that it was dead, but it did not stir.

Close examination showed the first shot had been directly between the eyes, through the mouth and into the upper gullet. The second shot had hit the nape of the neck breaking the spine. It was also now obvious, in examining its head, how many times he had been hit previously. Among the numerous injuries was a left foreleg that had been smashed by a bullet above the joint which had long since healed but had left the limb stiff and crippled. Gouges along his back showed where bullets had grazed him and left their marks in the form of deep ruts. We were now becoming more and more convinced that the monster actually was Shaitan and not just a normal, large mugger.

Cecil Pritchard sits on the killer croc, Shaitan, to whom twenty-four known human deaths were attributed.

We called to some of the men to help us skin the animal but, surprisingly, no one stepped forward. In fact, everyone retreated a step or two when we made the request for help. In the minds of the villagers there was no question that this was Shaitan who had terrorized them for years, and they were not about to jeopardize themselves by coming near the mugger and risk being possessed by the reincarnated souls of his many victims. Even two young men, who had been only too eager to help us skin and carry other crocodiles we had shot on earlier hunts in the area, showed no courage or interest in coming anywhere near the dreaded killer croc.

Realizing we could not obtain any help from our audience we had to undertake the difficult task of skinning the croc ourselves. Lacking proficiency, we struggled for about three hours removing the hide, leaving plenty of flesh on it to avoid making unintended cuts. The village people sat or stood at a respectable distance quietly watching but offering no help.

When the skinning was complete, we opened up the stomach of the dead beast. With a deft slice the stomach opened and revealed a veritable treasure trove of objects. One by one the gruesome relics of his former victims presented themselves. First there were four toe rings and a brass anklet. Then we pulled out a pair of brass bangles worn by young girls. The stomach also contained a small brass dish called a *thali*, which was used for ceremonial offerings. Everything was covered in a grisly green coating of cupric carbonate from the mugger's stomach acids.

We called some village people over to reclaim the objects but they shied away in utter horror. From a distance we could hear them talking among themselves as they identified each item we pulled from the stomach and named the person who had worn it in life before their fatal date with the mugger.

We tied the huge skin onto poles with manila rope and with extreme effort carried our bundle back to the rail station at Kataria. The weight, along with our rifles, made it virtually impossible to go more than 100 yards at a time before collapsing under our load.

When the freight train for Katihar West finally arrived, the station master kindly signalled it to stop, but the unobliging train guard, seeing our load and noting with obvious dissatisfaction how badly Cecil, Shaitan and I smelled, ordered us to an open car used for transporting sugar cane.

Arriving at Katihar West Station we located some more cooperative coolies who carried our mugger skin for us. In the shade of a tree we laid out the huge skin and, comparing it to a length of rope we had used to measure the monster on the river bank, we calculated a firm measurement of 19 feet 6 inches in length with a body width of 54 inches along the belly portion from the center of the side along to the mid-section. We never saved the upper back portion because such skin sections had no value. We consequently never obtained a full girth measurement.

In later years, World War II brought American soldiers to the area to be posted at Kursela Station and they had an opportunity to shoot some of the remaining muggers still lingering at the huge pool. None, however, equalled Shaitan in size or reputation. All are now gone and hunting of the huge muggers is a thing of the past which allows only reminiscences for those of us who were privileged to tread the outback areas in years past.

REFLECTIONS ON WHEN MAN IS HUNTED— BY MAN!

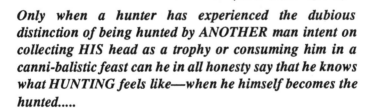

by

Capt. John H. Brandt

Only when a hunter has experienced the dubious distinction of being hunted by ANOTHER man intent on collecting HIS head as a trophy or consuming him in a canni-balistic feast can he in all honesty say that he knows what HUNTING feels like—when he himself becomes the hunted.....

I was sitting at a large round table at the headquarters of the Safari Club International surrounded by a group of many of the most prominent big-game hunters currently alive. It was a meeting of the Board and I was at that time serving as an officer of the organization. As each person spoke in rotation around the table describing his current organizational functions, the presentations were often interspersed with anecdotes of recent hunting adventures. As my turn came to speak, I made a comment about it being truly hair-raising to be charged by a wounded lion or to try to escape a rampaging elephant askari bull in dense bush. I added, however, that the ultimate reverse thrill in hunting was to have the tables turned and become the hunted! Adding to the presentation, I said that any inadvertent mistake during such an experience might result in *your* hanging on another trophy hunter's wall to be preserved as we maintain our taxidermy mounts—or worse yet, to suffer the indignity of being eaten by your adversary. I referred not to animal adversaries but to human head hunters and cannibals!

I had just returned from an eleven-year stint in the South Pacific and Southeast Asia. I was still undergoing re-entry crisis into American society and when my colleagues talked of organized safaris to Africa and the mountains of North America guided by competent professionals, my mind still revolved around tropical jungles, fetid swamps, leeches, tigers, king cobras—pygmies, cannibals, headhunters and opium growers in areas where I had just spent much of my adult life.

When the meeting was adjourned, one of the group approached me and said, " John, you were kidding about being the stalkee rather than the stalker weren't you? That sort of thing ended in the last century—didn't it?" Such an observation was indeed valid, although incorrect, when explorers and scientists were already probing outer space and the depths of the ocean floor. To realize that only within the past decade there were still corners of the globe which modern man had not yet reached seemed almost inconceivable. It was nevertheless true!

During my years in Asia I had experienced many opportunities to poke into such undiscovered corners. I had been an observer along the Sarawak-Kalimantan border in Borneo among the headhunting Iban during the Indonesia-Commonwealth border war.

I had collected for several American museums in New Guinea when permission to enter many areas labled "blank-unexplored" still existed. The Australian administration had graciously allowed me to go to areas where constabulary escort was still required and where showers of arrows often made visitors feel decidedly unwelcome!

At the same time in the Philippines, city dwellers in Manila would state emphatically, in embarassed astonshment, that headhunting certainly no longer existed in their islands. Yet the newspapers only twenty years ago regularly carried clips about the headhunting raids by Ilongots of Nueva Vizcaya in northern Luzon.

Newly pacified natives of the Jimmi River area of New Guinea pose with Brandt carrying huge spears tipped with cassowary bills and wrapped in cuscus fur.

In southeast Asia I had discovered and published on a previously uncontacted band of Semang pygmies in the forest of Satun Province, Thailand, and had contacted another band of unknown forest pygmies in Naratiwat Province on the Malay border before terrorists closed the areas to anyone who had illusions of returning alive. Who these people were remains a mystery to this day. In the north of Thailand, on the border with Laos, dwelled the *Phi Tong Luang*, the so-called Spirits of the Yellow Leaves, who had first been contacted in 1924 by an Austrian ethnologist and were not seen again until 1965. Their footprints in uninhabited jungle areas were occasionally observed and their abandoned lean-to's, covered with dried leaves from which they derived their name, were often seen by jungle hunters, but the people themselves remained illusive forest ghosts. North of the Golden Triangle, where the world's finest opium is grown, lived the headhunting Wha who were never subjugated by the British Raj and relied on collection of human heads to assure the fertility of the seed rice for next year's crop. During the early 1960's, Wha tribal elders, accompanied by an entourage of their finest warriors, marched from their jungled homeland in Burma to the northern Thai city of Chiengmai causing quite a stir as the feathered, heavily armed men carrying great chopping swords entered the city. The Wha had heard that the United States and Red China were having problems at that time and the Wha wanted it known that they also had a traditional enmity toward the Chinese. The startled Consul was shocked when the Wha offered a detachment of their best warriors, as a sign of good faith, to be used as the Consul wished, hoping for assistance in turn for their own private war against the Chinese in upper Burma and Yunnan. Although the offer was not accepted, I have often wondered what spectacular military use could have been made of as colorful a detachment as a group of sword-swinging Wha!

In south Thailand, Brandt discovered one previously unknown band of Semang pygmies and another that still remains to be contacted for the first time.

The diminutive Temiars of central Malaya often saw an RAF helicopter before they saw an auto. Armed with blowguns and poisoned darts, they played a unique role in the Malayan insurrection.

In Malaya, I had been involved with counter-insurgency work and had visited most of the deep jungle forts of Perak and Pahang in the central mountains. Here tribespeople, primarily Temiars and Semai, were being resettled close to forts in an effort to stamp out assistance from jungle dwellers to terrorist insurgents. The normally passive and unaggressive Temiars had been armed with single shot shotguns to protect their villages and to become "bounty hunters" by waylaying and ambushing known terrorists. For the tribesmen who possessed unsurpassed jungle savvy, the hunting of man became a most exciting new diversion! RAF helicopters had opened this area up so that a 30-minute flight would reach an area that I had previously taken two weeks to reach on foot. The helicopter was surely one of many revolutionary technological changes taking place in the area where people saw a chopper before they saw a car and a radio before they saw a light bulb; but only a recent "yesterday," less than a decade or two ago, the world still had not touched the realm of the jungle dwellers or had only dented their outer perimeter.

As a hunter, I had often wondered how it felt to actually be the hunted. How an animal must feel being trailed, ambushed and shot at. Assuming that animals do not normally think in an analytical way, the comparative corollary is perhaps inappropriate. I make a mental distinction between a carnivore that stalks man as a food source and becomes labled a "man-eater," and a rogue, be it elephant, boar, buffalo, or the like, that may kill people for whatever reason, provoked or not, and becomes a "man-killer." They are equally abhorrent and society makes every effort to kill or remove them as rapidly as possible for having the audacity to stand up to man! Society similarly finds headhunting and cannibalism equally unacceptable and initiates quick action necessary to stamp it out since it is unacceptable to us—but not to the perpetrators!

Most military combat veterans would have experiences where they could in all honesty say they have been stalked and felt like they were being hunted. This is certainly true, and the end result—death—is equally final. However, such experiences would be more comparable to territorial disputes by animals exercising their territorial imperatives to drive out interlopers, whether it involves a carnivore's hunting area or humans fighting about who has specific political jurisdiction over borders, islands, rivers or countries. Similarly, violent acts might be predicated by a bull evicting competitors for the females, often resulting in death for one or the other. It is easy to conjure up similar instances in our human experiences of death resulting from disputes of this

Headhunting was officially banned in New Guinea in the 1930's, yet many of the highland areas were not penetrated by patrols for the first time until as late as the 1970's.

kind. None of those, however, compare to a man being hunted for the specific purpose of his becoming a trophy to be preserved, revered and displayed. A final indignity might be that he is eaten by his slayer or that his body parts are used in ceremonial rituals. I suppose, at that stage, it would make little difference to me, although my wife often remarked that my shrunken head would be readily recognizable by my handlebar mustache. Ever after, it has received much more personal care and consideration!

Headhunting in New Guinea was officially banned in the 1930's. However, the Wahgi Highlands were not penetrated by explorers until the mid-1930's for the first time and much exploration did not occur until well after World War II. When I was in the area in 1959, the original explorers, the Leahy brothers, James Taylor and the Catholic Priest, Father William Ross, were all still young men. The first north-south traverse of Dutch New Guinea was completed that year and it was several more years before the Baliem Valley was first fully explored. Governor Nelson Rockefeller's son, Michael, was lost on the Casuarina Coast of southern New Guinea during the same period and it is speculated that he was killed by Asmat natives although his body was never recovered. Tribes of that area wear human skulls around their necks that are used as head rests, which I find hard to conceive as conducive to a restful night's slumber.

In northern New Guinea, the huge Tamborin houses of the Maprik Hills were built for religious ceremonies where human head trophies were prominently displayed. Women were strictly prohibited from seeing the interiors of such structures, or their contents, where men of the tribe communicated with the Gods. Although the area has been administered since German colonial times, prior to the 1914-1918 War, any new Tamborin house construction prompted questions in the minds of the administration. Sometimes it was better not to ask questions for fear of having the answers confirmed. Attacks on patrols in the

Jimmi River area and a disastrous raid at Telefomin on the upper Sepik occurred in the 1950's before the natives realized that fighting "city hall" was futile and that further efforts to perpetuate the old ways were coming to an end. The government had decreed that "thou shalt not kill" your enemies and chop off their heads, even if it is essential to establish your manhood, and that you may not eat your enemies no matter if done as a means of obtaining protein or with deepest reverence and respect for the slain.

The strange viral disease known as Kuru, endemic to cannibalistic tribes of the Lemari River, would also soon come to an end. At that time the unusual phenomenon, which afflicted mainly women and children of the Fore Tribe, was not fully understood until it was recognized that eating the flesh of enemy dead was done primarily by women who, in turn, gave scraps to children. Cannibalism follows many different patterns and Fore men did not usually indulge in such snacks. When man-eating came to a halt, the disease, the only one known to occur from the eating of human flesh, was soon to be relegated to the archives of medical research. Surprisingly, the slowly developing virus did not strike down some of its victims until the present time, almost thirty years later.

It was planned that overall penetration of the remote jungled mountain valleys of New Guinea would only be completed by 1972, a few years before Papua-New Guinea joined the world of independent nations. The same natives, who in 1959 fired arrows at me in defense of their homes from strange intruders totally beyond their frame of reference or comprehension, now run about with transistors, T-shirts and sunglasses to welcome tourists into some of the more accessible areas. At that time, while collecting for a museum, I had been constrained from buying a necklace made of human fingers by a concerned Australian Patrol Officer who said, "John, if they know they can sell that, you'll have a whole mob standing in front of your tent in the morning trying to sell you brand new ones with the blood still dripping out of them."

Values only take one generation to change. Yes, in the 1950's-1960's, headhunting and cannibalism was outlawed, but so is prostitution, drug use and gambling in our society! Time-honored ways disappear slowly.

Some years later, I had an opportunity (although I questioned the benefit at the time) to reach the upper stretches of the Rejang and Baleh Valleys of northern Borneo. Iban (Sea Dyak) longhouses were scattered along the wide jungle streams. British troops were operating in the frontier area in their border war with Indonesia. Ibans had been enlisted as invaluable scouts and guides. Ibans were among the most notorious of the world's

The ultimate hunting trophy! Human skulls examined by Brandt were prominently displayed in baskets hung in Dyak longhouses in Borneo.

headhunting societies. As opposed to neighboring tribes such as Kayans and Kenyahs, who also took heads as trophies, Ibans alone believed in quantity as opposed to quality.

Borneo's Dyak tribesmen were among the most notorious headhunters in Asia. In the late1960's extensive headhunting raids were still widely reported by international news services.

During the 1930's in the reign of Rajah Brooks, the so-called white Rajah of Sarawak, headhunting had been banned, but shortly the Japanese invaded during World War II and the restriction, intentionally or not, was lifted. In the longhouses, baskets of human heads adorn the entry way into each warrior's cubicle and many of the skulls possess beautiful gold teeth attesting to the quality of aesthetic dentistry of the Japanese. Ibans would delight in taking down the head baskets and would describe in intimate detail how each victim was killed, what sort of an adversary he was and when and where he died. Sounds very similar to someone taking a visitor on a tour of the trophy room to see our moose, bongo or dik-dik doesn't it? Iban fingers are tattooed to indicate successful head-takings. Even in primitive societies, in the absence of Record Books, such events must be properly recorded!

As during World War II, headhunting was still prohibited officially, but the exigencies of war permitted temporary relaxation of the laws. Twenty-nine heads were collected from border raiders during my time in Borneo, although officially such things were not supposed to be going on. In December 1967, international news services shocked the world by describing head raids conducted by Dyak tribesmen. Border insurgents had aroused the Dyaks' anger by burning some villages and the tribesmen in retaliation had passed the blood cup, a traditional signal for war, to assure allegiance of neighboring longhouses in the counterattack. Befeathered, blood-streaked, sword-wielding warriors went forth attacking initially known Chinese insurgent communities; but with the blood lust running high, this soon disintegrated into full scale attacks on any Chinese in the area. Deaths totaled into the hundreds and some 50,000 Chinese were forced into refugee camps on the coast to escape the wrath of the headhunting Dyaks. Multitudes of grisly new trophies soon adorned the hunters' longhouses. A person killed in a raid would serve little purpose if his head were not brought home as a trophy to be viewed and admired by the people of the longhouses.

I have sat as an honored guest in Iban longhouses in *Bedara* ceremonies where my Iban hostess, beautifully bedecked, would sing me little ditties which translated into, "May the women of your village hear you coming by listening to the splashing of blood on the trail from the basket of human heads you will carry home . . . ." Sweet, well-intentioned sentiments from a young woman who could conceive of no other way for a man to distinguish himself and prove himself worthy of her interest.

Among some tribes of Asia, headhunting is not a constant or necessarily a military pursuit. Often, it is believed that only the drippings from a human head can assure the fertility of the seed rice. In such cases an elaborate ambush is planned by trickling a handful of last year's rice across a trail and then awaiting the arrival of a suitable victim. The intent is that the very last thing the victim is to see before his impending death is the rice sprinkled on the trail. As his mind comprehends what is to unfold, before he can say, "Whoops, I'm in deep trouble," his head is removed from his body. It is ritually hung over the bowls of seed rice accompanied by appropriate ceremonies. The primitive hill farmers see nothing wrong in this murder and cannot run the risk of not properly assuring the fertility of the seed for next year's crop.

Because of these seasonal killings, often bandit groups or other jungle opportunists take advantage of such ambushes to carry out robberies or revenge killings of their own. It is only necessary to remove the head and the hill tribes will be blamed. Differentiating between the two can cause endless frustration for police administrators.

The taking of human heads is a very specialized art requiring special weapons of unique manufacture. In the Philippines an axe is used with a curved cutting edge counterbalanced with a pike at the upper end of the axe blade. In Borneo a sword is used with a concave-convex cutting blade so that, depending on whether the chopper is right-handed or left-handed, the blade will always deftly slice into the body. A special, short-bladed knife is placed in the sheath which is used to surgically remove the head and retrieve the axis bone, where the head joins the backbone, which is strung onto a belt for carrying the sword.

In Melanesia, Fijians developed a special multi-pronged fork for eating human flesh—one of the few forks ever developed in the primitive world. Among the KuKuKuku of New Guinea, specialized drainage conduits were placed below bodies laid on scaffolds so fluids dripping from the cadaver could be collected and consumed.

Only in Asia? Hardly. Cannibalism existed in many areas of central Africa and the technique of shrinking human heads practiced by the Jivaro of the upper Amazon is unique and well-known. Only quite recently, rules of law penetrated the Jivaro homeland and prohibited such practices.

Cannibalism probably reached its most celebrated heights in the Valley of Mexico where only recently anthropologists have reviewed the literature and descriptions of Aztec ceremonial practices. Long considered as sacrifices to various gods, the question constantly remained about why so many sacrifices were necessary and why the populace relished such continuous bloodshed. Decriptions by the Conquistadors of Cortez of the dismemberment of sacrificial victims, after their hearts had been removed in a ritual excision, coupled with the fact that thousands of skulls lined vast collecting depots for the bodies, were earlier attributed to propaganda to make the Aztec enemy appear savage, barbaric and very much in need of subjugation by the invading Spaniards. Now more analytical scholars are recognizing that the densely-populated agricultural communities of central Mexico raised crops deficient in protein and had no domestic animals nor large game populations that they could obtain meat from.

Gruesome as it now appears, strong evidence exists that the sacrificial victims were intended from the beginning to end up as stew and that connecting the killings with religious justification only made the practice more acceptable. Warriors, as a consequence, were encouraged to bring home prisoners which would provide consumable victims for the knife-wielding priests on the pyramids. It also explained the Aztecs' aggressive conquering of surrounding areas but rarely absorbing these into the Aztec political superstructure. If cannibalism is accepted as a means to an end, then having neighbors as enemies, which could be gathered and slain, made good sense. One can eat a nondescript, nameless cow far more easily than a pet dog, cat or goldfish!

Have I ever been stalked by a man-hunting trophy seeker? Obviously, not successfully, although I lived for six years where these things were going on. But by comparison, how many hunters know if they have ever been looked over as a potential victim by a man-eating tiger, lion or leopard? If the attack is not made, one is unaware that the danger ever existed. Does any hunter walking along a river bank or wading through a swamp in Africa, Asia or South America know if a crocodile or alligator lurked only a few feet away, but decided not to attack for reasons known only to the predator. The victim remains blissfully unaware!

As the world changes, so do the practices described. More danger from other human predators quite likely now exists walking down lonely, dark streets of most metropolitan cities than walking down trails in the jungles of distant lands. Headhunting and cannibalism are now rare even if not totally eradicated, just as the total trade in the Middle East in human slaves has not ended after a century of world effort. Yes, changes have occurred, but the past was only yesterday, and the changes only the day before!